AN
INTRODUCTION TO
MEASURATION
AND CALIBRATION

An Introduction to Measuration and Calibration

by Paul D.Q. Campbell

INDUSTRIAL PRESS INC.
New York

Library of Congress Cataloging-in-Publication Data
Campbell, Paul D. Q., 1959–
    An introduction to measuration and calibration /
by Paul D.Q. Campbell. — 1st ed.
        208 p.   15.6 × 23.5 cm.
    Includes index.
    ISBN 0-8311-3060-1
    1. Physical measurements. 2. Calibration.
3. Metric system. 4. Weights and measures.
I. Title.
QC39.C324 1995
530′.7—dc20   94-31185
                                                        CIP

INDUSTRIAL PRESS INC.
200 Madison Avenue
New York, New York 10016-4078
AN INTRODUCTION TO MEASUREMENT AND CALIBRATION
First Edition
1995

3 4 5 6 7 8 9 10

Dedication

To Beverly

My loving wife, my best friend, and my toughest critic.

For untold countless hours of thankless labor and devotion to projects beyond your control, without a whimper. I greatly thank you.

Contents

Acknowledgements

THIS IS ALWAYS MY FAVORITE PART of the book to write, perhaps because it is one of the last in any book I write. It's also because I have an opportunity to put into permanent record the contributions of those who don't get their names on the cover.

First and foremost, I would like to express my warmest thanks and blessing to Leslie C. Sultz and Diann Sultz for keeping me alive, quite literally, and sane, while I hammered away at this text. Without their help, I would never have finished.

I would also like to express my thanks, and especially my admiration, to my friend Woody Chapman. May we *always* be friends. Thanks also to Lisa Chapman, for tolerating all of us in this business.

I could not forget my many friends at Industrial Press. John, thanks again, everything is great, I look forward to hearing from you again; I mean, I REALLY look forward to hearing from you again. Steve, everything looks great, and thanks for making me look good. I especially owe thanks to Carol (as all of us at I.P. do) for keeping everything connected. Without her efforts, it might be difficult to keep track of me at times; and her cheerfulness is perpetual.

Thanks are also in order to my parents, Robert A. Campbell and Pauline A. Campbell, for individual efforts too numerous to list. Suffice it to say they have labored physically, mentally, and emotionally to see that I completed this book. Thanks also to my mother-in-law, Mary Gerlach, for her enthusiasm.

A cursory glance through this text will quickly tell you that I am also deeply indebted to many companies and individuals for their generous support of materials and information.

Barbara Pralinsky, of The L.S. Starrett Company, has been a great help in securing photographic support and information for this text (and other texts like it, including my own and, I am sure, those of other authors), without so much as a card on Sweetest Day. She is a credit to The L.S. Starrett Co., to the publishing and manufacturing businesses, and to herself. Many thanks from myself and my wife.

My thanks also to Michael McCue of The L.S. Starrett Co., Florette Sinofsky and DoALL Co., Howard Brown and Mitutoyo/ MTI Corp., J. Bruce Whyte and Federal Products Co., Walter Tress and Cubic Precision Corp., Zygo Corp., Bill Otis and J & L Metrology, Cathy Powell and Fanuc Robotics, Michael A. Lavey and Stocker & Yale, Inc., and Carr-Lane Manufacturing.

Finally, my thanks to Gerry and Patty Holt, Joe and Devin Brown, Joe Bohr, and everyone else who either offered advice, support, or in general had to tolerate me while I laughed, cried, moaned, rejoiced, and meandered around the country.

Introduction

THE GREEK PHILOSOPHER ZENO proposed the idea that between any two points in space there is a distance divisible by an infinite number of progressively smaller units. His contention was that it is impossible to traverse an infinite number of units, and therefore impossible to know the distance between any two points. This became one of the now famous "Zeno's Paradoxes."

This was—in its time—interesting, but certainly not of any practical use to an industrious society bent on knowing and controlling its environment. While mathematics and mensuration are not mutually exclusive, they were born out of different ways of thinking. Mathematics is the progeny of philosophy, with abstract concepts such as infinity and set theory. Mensuration was born of practicality and a need to do business.

(It should be noted that *technically* there really is no such word as "measuration." The term in the English language is actually "mensuration." However, I have opted to use the colloquialism measuration because it is more common in usage and generally easier to pronounce. This is only mentioned so that the astute student of the English language will not perceive an error and thus read the remainder of the book with suspicion.)

Measuration quickly evolved from the need to translate any measurement into a standard unit, with standard subunits, that was easily understood and accepted by everyone who would have cause to use it. In its simplest form, it is a means of communicating information about the physical properties of an object. Roughly 6000 years ago, mankind came to the realization that between any two points there existed a distance, and the field of measurement was born. Like most human inventions, measuration was born out of necessity—specifically, the necessity to record and relate information about products, places, or parcels of land. At its inception, measuration was based on the measuring instruments most readily available to everyone—the parts of the body. This is still largely discernible from the names of some of the units of measure, e.g., foot, hand, and span. Other units still in use today are also based on these body-part measurements, although their names may not make it so obvious, e.g., yard, cubit, and inch. The yard was the distance from the end of the nose to the tip of the middle finger; the inch was the distance across the width of the thumb; and the cubit, although infrequently used today, was the distance from the bent elbow to the tip of the middle finger, or approximately one-half of a yard.

All of these units of measurement obviously had varying degrees of accuracy associated with their relative size, and likewise had different applications. The one common factor they did have was a serious lack of conformity to a given standard. My hand and my neighbor's hand are substantially different sizes. Therefore, if I were going to buy a horse from her (incidentally, horses are still measured in "hands" to this day), she might be able to claim the animal was 17 hands high, but I would measure it as only 14 hands high. Both might be correct, but they are based on different standards of measurement for the same unit. The system went through various refinements: the yard, for example, ceased to be the distance from the end of "your" nose to the end of "your" middle finger, and eventually was given the standard of the distance from the thumb tip to the nose on King Henry I. This was still not terribly accurate, but it had the redeeming quality of a consistent standard, at least while King Henry I was alive, and one that was not open for a great deal of dispute. Of course, King Henry's mortality would eventually catch up with him, so a more permanent standard was developed: the distance between lines inscribed on two gold inserts in a bar made of bronze was measured at a particular room temperature to compensate for expansion or contraction of the metals.

Accuracy and consistency obviously have taken a quantum leap forward with the advent of this bar, but change and striving for precision are the hallmarks of technological advance, and the standard would again need to be made more accurate. This time, the length of a lightwave from a particular color and type of light was denoted as the standard on which all forms of linear measurement would be based. To date, this has proven to be sufficiently stable as a standard.

However, it should be noted that this standard was not possible until science 1) understood that light traveled in waves, 2) developed a means of measuring these waves, and 3) determined the stability of these waves. This method of measuring lightwaves, called interferometry, is used in some precision measuring devices which we will cover in detail later in this book.

The development of measurement standards has been refined into two predominant systems—the Metric System and the English System. For years, a heated debate has gone on both publicly and privately about the adoption of the Metric System by everyone on the globe. To date, this has not transpired, and it does not appear it will soon. Therefore, students of measuration will need to be completely familiar with both systems in order to deal with any problem placed before them. To that end, we will first cover in detail the two measurement systems, and throughout the book we will attempt to use examples of both. In a world of shrinking global manufacturing borders, the engineer, scientist, and machinist will need to be able to read and properly interpret information from sources that may not use the same units of measurement they are accustomed to using.

An
Introduction to
Measuration
and Calibration

Chapter 1

Measuration Systems

A MEASUREMENT SYSTEM is, in essence, a "standard." A standard for our purposes is an established rule, or benchmark, by which information is judged. This means that when the distance between two points is specified in a particular number of units, that distance will always mean the same thing to all who encounter it. Unfortunately, the problem with most "standards" is that they are only standard to a select group who choose to use them.

Measurement standards are largely country dependent. The "English System" is used essentially by the United States and a few other nations, and the "Metric System" is used by most of the rest of the world. Without being dragged into the issue of "Why doesn't the U.S. just adopt the Metric System?" suffice it to say that it is a problem since the one nation best known for using the English System is also one of the world's largest industrial manufacturers. However, we are not political activists, we are scientists, and therefore we will learn both systems and leave the debate to the politicians.

It is absolutely imperative that the engineer or machinist be "bilingual" in the science of measurement. For example, while the company you work for may design and build products using the Metric System, the material supplier will very probably manufacture and stock materials designated with English units. Conversion from one system's units to the other's will have to become second nature. While precise conversions may be done with an electronic calculator or computer, the engineer or machinist should be familiar enough with the conversion factors to make close approximations from memory.

There is one bright spot in all this: while linear measurements may be in either of the two predominant systems, angular measurements will always be the same. Now for the bad news: there are three different units of angular measurement considered universal and completely independent of a system standard. Typically, angles are designated in "degrees," but you should at least be familiar with the remaining two types of units in the event that a conversion must be done.

The English System

It is actually odd that the English System is called the English System, since it was not developed in England, and with only a few exceptions is not even used there today. The English System is the system developed from the size of the parts of the body, as we discussed in the Introduction. For example, the "foot" was considered to actually be the length of a human foot.

The English System works as well as any system. There is only one thing inherently confusing about it—it uses either decimal or fractional subunits. This becomes even more complicated since the decimal translation in feet is radically different from the decimal translation in inches. The only saving grace of this system is that the fractional subunits are always created by halving the base unit, i.e., one-half, one-quarter, one-eighth, one-sixteenth, and so on. If other bastard denominators were used, the mathematics involving two or more dimensions would become nightmarish. Imagine trying to find the lowest common denominator for 1/25 and 1/27 every time you needed to add two linear dimensions.

By simply doubling the denominator or, in other words, halving the subunit, fractional dimensions can be mathematically manipulated with comparative ease. For the sake of clarity, we are going to proceed on the premise that the "inch" is the standard unit of measurement of the English System. Virtually every

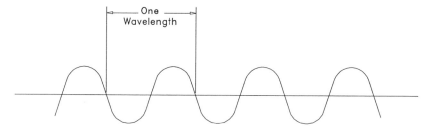

Fig. 1-1. The current inch and meter are both based on a multiple of wavelengths of monochromatic light. A wavelength of light is measured from the middle, or mean, of the wave form to an identical place on the next wave form.

other unit in the English System is definable in terms of the inch, since it is one of the smaller units of the system, and smaller units are generally some even decimal division of the inch.

For example, the "mil" is a unit of length in the English System which is equal to one one-thousandth of an inch. It is an even numbered decimal subdivision of the inch, but would be much too small to use as a standard. This term, in particular, should be used with caution. The *term* mil is technically one one-thousandth of an inch and is typically used to describe the linear thickness of coatings and films. However, the *expression* mil is often used as a colloquialism for "millimeter" which is a unit of the Metric System, and a radically different distance.

So just how big is an inch? We discussed in the Introduction that the inch was originally considered the width of the human thumb across the first joint. Now, like most unit standards, it is based on the length of a number of lightwaves. In 1960, the International Bureau of Weights and Measures established the standard of the inch as being equal to 42,016.807 wavelengths of the monochromatic light from Krypton-86 gas. Fig. 1-1 illustrates how a wavelength is determined. We will cover in detail how to read linear measurements on the steel rule for both the English and Metric Systems at the beginning of the next chapter.

Metric Dimensioning

Please don't skip this section because you either don't like metric dimensioning or don't understand it. Like it or not, if you plan to work in industry, you will need to know the Metric System of measurement.

A little background is in order. Our customary system of linear measurement is commonly referred to as the English System. Most of us find it easy enough because it was the first one we were taught as children.

Somewhere along the line, someone decided that a more stable unit of measurement was in order, perhaps because the width of a thumb varied so much from person to person. At any rate, they set about looking for something of fairly consistent length. At one point, the distance across the flats of a honey bee's comb was actually considered, but later that was also abandoned.

Finally, in 1790, the size of the Earth itself was settled on as a pretty stable size. Being too big to be practical, the distance of a line from the equator to the north pole through Paris, France, was divided into a reasonable number of units (ten million) to produce a single unit of practical length. Thus the "meter" was born.

However, the Earth itself even proved to be lacking the homeostasis required by some nitpickers, but the meter had been around long enough by then so that no one was willing to abandon it along with the poor inch. So in 1960 the meter was redefined as being equal to 1,650,763.73 wavelengths in a vacuum of the orange-red radiation of Krypton-86. We now have a single unit, of extremely stable length; the only thing left is to divide it into some reasonable type of subunits.

When the English System was developed, it seemed reasonable that any unit could simply be divided in half to obtain smaller units. Each subunit could then be divided in half again and again ad infinitum to obtain the desired degree of accuracy. However, mathematical calculations involving 27/64 and 7/8 became a nightmare which plagues fifth graders to this day. With the advent of the electronic calculator, conversion to decimal fractions became a lot quicker, but you still ended up with a result resembling the value of Pi.

Not wanting to be faced with this terrible quandary in the sixteenth century, it was decided that the meter would be divided by ten. Each subunit created by this division was in turn divided by ten to reach the degree of accuracy required. Now instead of worrying about lowest common denominators and

Metric Prefixes	Symbol	Multiples and submultiples
Tera	T	1,000,000,000,000
Giga	G	1,000,000,000
Mega	M	1,000,000
Kilo	k	1,000
Hecto	h	100
Deca	da	10
Deci	d	.1
Centi	c	.01
Milli	m	.001
Micro	μ	.000001
(greek mu)		
Nano	n	.000000001
Pico	p	.000000000001
Femto	f	.000000000000001
Atto	a	.000000000000000001

Fig. 1-2. The Metric System employs a series of prefixes to any unit to indicate the multiple or submultiple of ten to which the unit is multiplied. Here we see that an attometer would in fact be a very short distance, and the terameter would be a really long hike (one trillion meters, or 62,137 miles).

converting back and forth from decimal, the decimal placement was built right into the measurement system itself.

Each subdivision or combination of the meter is named by adding a Latin prefix to the word "meter." This method was undoubtedly chosen for the sake of brevity. The Latin prefix "milli" means "one thousandth of." Therefore, millimeter is simply "one thousandth of a meter," but is much easier to say and write.

Fig. 1-2 gives the Latin prefixes for the metric system and their equivalent amount of the associated unit. In our case, the equivalent unit is the linear measurement of the meter, but the same prefixes are used for weight (grams), liquid (liter), and other measurements. For a more detailed explanation of other metric units, refer to Appendix A.

Appendix A gives a conversion table of fractional inch, decimal inch, and metric distances in millimeters up to one inch.

However, you will frequently encounter bastard numerical distances, which you may need to convert to or from metric. Bastard numbers are considered to be those numbers which are not some even division of an inch. Even divisions include 1/2, 1/4, 1/8, 1/16, 1/32, 1/64, and their associated decimal conversions. Bastard numbers might be 1/3, 1/5, 25/27, 0.2285, 3.14159, and 1.41421. A detailed discussion of converting English and Metric dimensions is also covered in Appendix A.

Angularity

Defining an angle is somewhat more involved than defining a line or linear distance. An angle is, by definition, the area between two converging lines, but it makes no provisions for the length of those lines. Therefore, an angle must be able to be defined regardless of the length of the lines forming it. To accomplish this, an arc is projected between the two lines emanating from the intersection, or "vertex," of the two lines. This arc is then divided into an equal number of segments. The length of the segment is immaterial for the most part, only the number of segments along the arc is relevant.

DEGREES

The most common segment division for the arc between two lines is called a "degree." The degree is equal to 1/360 of a complete circle. This means that when our two lines forming an angle are projected onto a circle, with the vertex of the angle exactly on the vertex of the circle, the arc formed between the lines can be interpreted by the number of subdivisions for the circle. Fig. 1-3 illustrates this point.

Most of us learned this procedure in school, but may have forgotten how the degree itself is subdivided. The subdivisions of the degree use the same names and similar principles as those used for subdividing units of time. Each degree in turn uses a subunit called a "minute," and each minute is subdivided into "seconds." Just as there are 60 minutes in an hour, so are there 60 minutes in a degree. Likewise, each minute of angularity is subdivided into 60 seconds.

Adding and subtracting values in degrees, minutes, and sec-

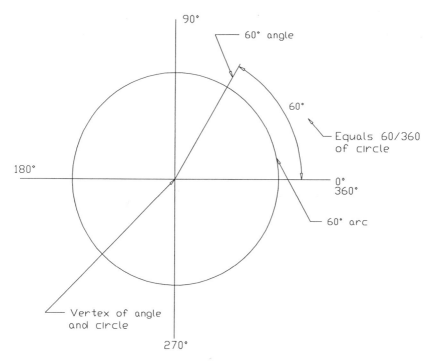

Fig. 1-3. Angles measured in degrees use a system based on the angle's relationship to the segment of a circle. The degree is equal to 1/360 of a complete circumference; therefore, any angle laid over a circle will cover a certain number of these subdivisions, or degrees.

onds is fairly straightforward, by simple columnar method, with any value in excess of 60 carried over to the next group. However, since many calculations involve electronics and thus decimal returns, it is important to know how to convert decimal degrees to and from minutes and seconds. This is accomplished by simply dividing or multiplying the decimal figure by 60. Each time this operation is performed, the integer part of the figure is set aside and labeled. The symbol for the degrees is a small circle to the right of the value, e.g., 360°. Minutes are designated by a single quote (or apostrophe), e.g., 359° 15′. Seconds are designated by a double quotation mark, e.g., 359° 15′ 37″.

If the value is given in decimal form, such as 36.8275°, conversion to degrees, minutes, and seconds will start with setting the integer aside and labeling it, i.e., 36°. The remainder, or decimal part, of the value is now multiplied by 60, i.e., 0.8275 ×

60 = 49.65′. Again, the integer is set aside and labeled along with the other labeled integer as 36° 49′. Again, the remainder is multiplied by 60, i.e., 0.65 × 60 = 39″. This value is now set with the others and labeled as 36° 49′ 39″. Suppose, however, that the decimal degree value had yielded something that was not an even multiple of 60, like 36.8386. Using the same procedure, this yields 36° 50′ 18.96″. Now we are faced with decimal seconds. This is perfectly acceptable, as there are no angular units in this system smaller than the second. Conversion from degrees, minutes, and seconds to decimal degrees is accomplished in exactly the opposite manner. Both the minutes and seconds are *divided* by 60, and the results added together. For example, to convert 47° 13′ 27″ to decimal degrees, first set aside the degrees, then divide the minutes by 60, i.e., 13 ÷ 60 = 0.2167 (rounded to four places); then divide the seconds by 60, i.e., 27 ÷ 60 = 0.4500; and then add the results: 47 + 0.2167 + 0.4500 = 47.6667°.

47.2242°

RADIANS

Radians are the second most common means of specifying angularity, and are typically used in mathematical computations involving angles. The radian is equal to an arc which is the same length as the radius it is projected from. For example, if the radius of the arc is one inch, then the length of an arc equal to one radian is one inch. Fig. 1-4 illustrates the radian. Notice in this illustration that as the radius of the arc increases, so does the length of the arc, but the angle remains the same.

The radian will always be equal to 57.2958°. Therefore, to convert radians to degrees, the radian value is multiplied by 57.2958. To convert degrees to radians, the degree value is divided by 57.2958. Since radians have no subunits, any remainder left during conversion is simply expressed in decimal form, such as 1.6875 radians. The word radian should always be spelled out to avoid confusion with the symbol "r" used to indicate radius. It is also necessary to convert degrees, minutes, and seconds values to decimal degrees before converting to radians.

GRADS

Grads are fairly uncommon, but are occasionally used in engineering computations, so they are at least worth understand-

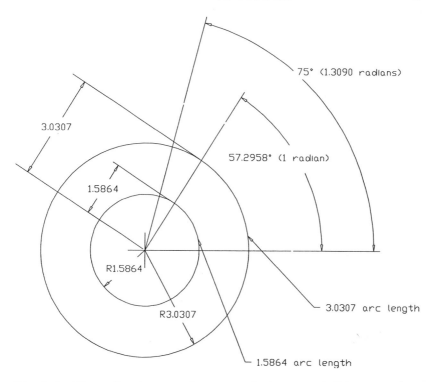

Fig. 1-4. The radian is an angle or arc which is equal in length to the radius of the arc formed when the angle is overlaid on a circle with a common vortex. In this example, we see that as the radius increases, so does the arc length, but the angle remains the same. The larger angle shown has the same radii shown, but since the angle is larger and the arc length formed exceeds the radius, it is greater than one radian.

ing. The grad is very similar to the degree, with two exceptions. First, there are 400 grads in a complete circle instead of 360 as there are with degrees. Second, grads have no subunits; therefore, grads will always be expressed in decimal form.

Conversion is merely a matter of multiplying the value by a constant. When converting grads to degrees, multiply by 0.9 (360 ÷ 400 = 0.9). When converting degrees to grads, multiply by 1.1111 (400 ÷ 360 = 1.1111). If we use the information already covered, we can determine that there are 63.6620 grads in each radian. Conversion between grads and radians can be accomplished in the same manner as conversion between degrees and radians, only using 63.6620 as the conversion factor.

Limits

Regardless of the measurement system used, most industrial measurements fall within a range of "tolerance," or "limits." This means that a part is specified to be a certain size, within reason. There is a certain amount that the part may be under or over size and still function properly. That amount is the part dimension's tolerance, and the difference between the upper tolerance and the lower tolerance is called the limits.

Typically, limits are thought of as a linear distance, but it is also true that there are angular limits. Limits are specified in a number of ways, and may be either unilateral or bilateral limits. Unilateral limits are specified by a numerical value for the part with a degree of tolerance that is only greater or lesser. Bilateral limits are specified by a numerical value for the part dimension with a degree of tolerance that is *both* greater *and* lesser. Fig. 1-5 illustrates the difference between these types of limits.

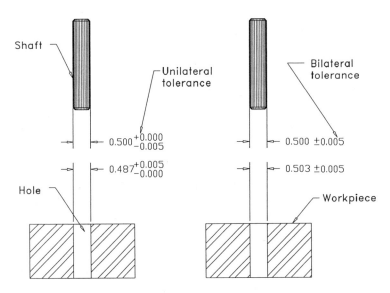

Fig. 1-5. Unilateral limits are those which are only greater or lesser than the dimension they are attached to, while bilateral limits can be either greater or lesser than the associated dimension. Typically, unilateral limits are used for parts that fit together with a high degree of accuracy, while bilateral limits are used for parts with a lower degree of accuracy.

This is significant because the reason for measuring many parts is not just to ascertain their size, but to ascertain if that size falls within the specified limits. Keep in mind that most parts are measured under two different sets of circumstances: first to make the part to the specified size, and second to check that the part has been made correctly. Checking the measurements of a part will generally employ instruments which are far more accurate than those used for making the part. While this sounds, on the surface, as though it places the builder at an unfair disadvantage to the checker, this is not actually the case. Tools to make a part need only be accurate enough to make the part, but tools to check a part must be more accurate than the minimum in order to avoid undetected errors beyond the capabilities of lesser instruments.

A general rule of thumb is that any instrument used for checking a part should be ten times as accurate as the requirements of the part. In other words, if the part must meet a tolerance limit measured in one-thousandths of an inch, then the measuring instrument should be accurate to one ten-thousandth of an inch. This is, of course, only a "general rule" not an axiom. For example, if the part has a tolerance limit of plus or minus five ten-thousandths of an inch, an instrument accurate to one one-thousandth of an inch is probably acceptable. If the parts must be accurate to one ten-thousandth of an inch, it is unwise to use an instrument accurate to only one ten-thousandth of an inch—not impossible, just unwise.

Linear Measuring: The Steel Rule

There are a plethora of different measuring instruments, some differing in function, but many differing in the degree of accuracy required. Those with the least degree of accuracy are generally also the most frequently used. However, being the "least" accurate does not mean they are inaccurate. It is possible to draw an analogy between measuring instruments and microscopes. The standard light microscope is the most common type used in many laboratories, but it is obviously not as accurate as an electron microscope. Yet the standard light microscope is much more accurate than a magnifying glass. For most applications, the light microscope is sufficiently powerful; and in some applications, the electron microscope is simply too powerful.

Industrial measuring instruments follow a similar pattern. The most common and frequently used instrument in manufacturing is the steel rule. It is not as accurate as a micrometer, but it doesn't need to be. The steel rule is a flat stainless steel strip that has the graduations of a measurement system etched into it. Each unit, whether English or Metric, is also divided into smaller subunits which vary from one rule to the next, depending on the given accuracy of the tool. Fig. 2-1 illustrates the steel rule.

Fig. 2-1. The steel rule is the workhorse of measuration instruments used in industry. They are available in various lengths from the six- and twelve-inch lengths pictured here, to steel rules six feet or more long. (Photo courtesy of Mitutoyo/MTI Corp.)

Chapter 2

Mechanical Measuring Devices: Linear

THE MOST COMMON MEASURING INSTRUMENTS are mechanical measuring devices. However, "mechanical" is not necessarily indicative of a "mechanism." For our purposes, mechanical measuring devices are those which do not employ light, optics, electronics (solely), pneumatics, or other secondary methods. A device that an operator reads visually does not make it an optical measuring device, i.e., reading the numbers on a ruler does not make it an optical measuring device, it is a mechanical measuring device.

Many mechanical measuring instruments do in fact employ a mechanism. Typically this is some means of opening and closing the instrument over a given distance, and holding it accurately in that position while the operator reads the measurement. Other nonmechanized, but still mechanical, measuring instruments such as many gauges simply either fit a part if it is correct or don't fit if the part is not correct. Hybrid instruments, which might be referred to as "electro-mechanical," will be considered electronic instruments for the most part and covered under the respective chapter.

The nature and thickness of the material this instrument is made of keeps it relatively stable from expansion and contraction due to temperature variations in the workplace. The instrument itself is analogous to a ruler used in elementary school or a carpenter's flexible steel tape line. However, not only is this instrument most stable in size, the graduations are typically much finer, and the steel rule is a rigid instrument. Generally, the steel rule is used when making parts, not for checking them. The steel rule has each unit of the measurement system designated on it in numerical succession. However, the subunits are only designated by a "tick mark" at appropriate intervals between the main units. The tick mark may or may not vary in length to indicate the different sizes of the graduation. To complicate this a little more, the graduations may be either fractional or decimal. It takes a little examination when working with any new steel rule to be sure of reading the proper amount of the graduations.

Let's examine a few of the different graduation techniques for a typical flat steel rule. Fig. 2-2 is an illustration of an English System steel rule graduated in inches and fractions down to 1/32 of an inch. Note that with fractional graduations the tick marks are shorter as the increment becomes smaller. This is advantageous in making accurate readings of the fraction with the correct denominator. This illustration denotes how each graduation on this type of rule is specified. There are several variations of this type of steel rule. One variation would have a still smaller and finer set of graduations at 1/64 of an inch. Another would have a second graduation set, or even another measurement system, on the opposing edge of the rule.

Alternate graduations are included for a couple of reasons. First, they might be added in larger increments to aid the eye in faster reading of the rule. Second, a completely different graduation set may be added to increase the versatility of the rule and eliminate the need for maintaining a second instrument. For example, one edge of the scale may be graduated in fractions, and the opposite edge could be graduated in decimal portions of the inch. Fig. 2-3 illustrates a double-edged rule for quick reading.

Another example of the steel rule is shown in Fig. 2-4 with a decimal graduation set. While these graduations are technically 1/50 of an inch, the conversion automatically makes them in graduations of 0.02 of an inch. Some rules are graduated in

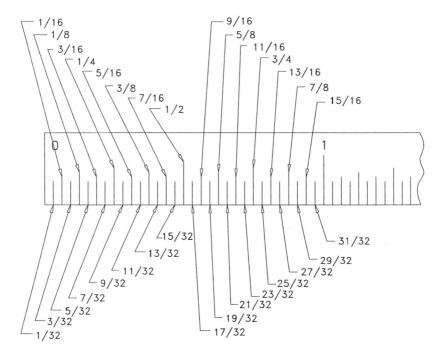

Fig. 2-2. The English System steel rule graduated in fractions utilizes progressively shorter tick marks for progressively smaller subunits of the inch for ease of reading. Each fraction is exactly one-half of the next larger subunit, so the denominator of each progressively smaller subunit is doubled.

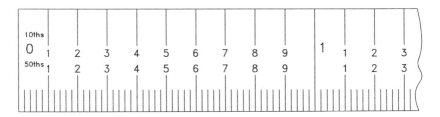

Fig. 2-3. Double-edged steel rules sometimes utilize identical measurement systems with different degrees of accuracy to facilitate easy reading. This particular rule is graduated in tenths of an inch on one edge and fiftieths of an inch on the remaining edge. The steel rules in Fig. 2-1 are also graduated on both edges.

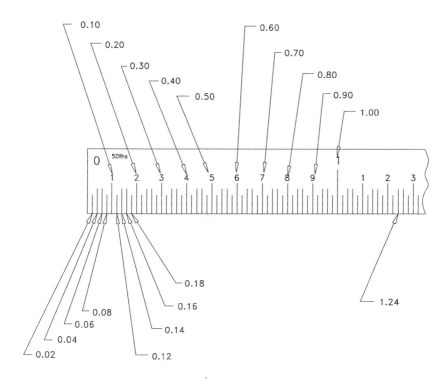

Fig. 2-4. Steel rules with graduations of one-fiftieth of an inch. This type of scale is especially well suited to making measurements in decimals of the inch since each graduation is 0.02 inch.

1/100 of an inch, or 0.01, but these are both costly and difficult to read. When accuracy beyond 0.02 of an inch is required, it may be advisable to select a different type of measuring instrument.

Metric steel rules have similar sorts of combinations. Fig. 2-5 illustrates a typical metric steel rule. Metric steel rules are graduated in millimeters or half-millimeters, with a designation at each centimeter. These rules range in length from about three decimeters long to meter and longer steel rules. Since steel rules can be graduated on both edges and both sides, it is common to find many steel rules with fractional inch and decimal inch graduation on opposite edges of one side, and millimeter and half-millimeter graduation on opposite edges of the reverse side. This combination makes one instrument far more versatile, and therefore overall tool costs are much lower.

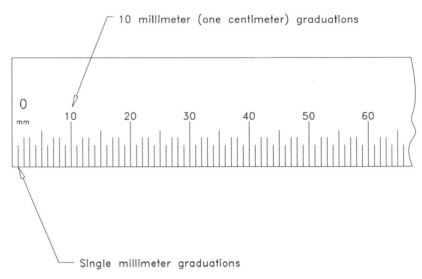

Fig. 2-5. The metric steel rule is graduated in millimeters, or occasionally half-millimeters, with designations etched every ten millimeters (one centimeter). Individual graduations are made different lengths, with the five millimeter tick mark a little longer, to facilitate ease of reading the instrument.

The Linear Vernier

As the demand for finer accuracy becomes greater, mechanical measuring instruments that are true mechanisms come into play. With the advent of electronics, many of these instruments now are available with a digital readout, but most are still manufactured with mechanical means of making the measurement. The completely mechanical instruments are just as accurate and far less costly to purchase and maintain. Therefore, it is important to know how to read these purely mechanical measuring devices.

The means that these instruments use to make such fine measurements is a system called a "Vernier." This system was developed and named for the seventeenth century French mathematician Pierre Vernier (1580–1637). The Vernier uses two opposing sets of graduations with slightly different tick marks over a given span from one set to the opposing set. This system may be placed with the two opposing sets of tick marks linear, on a set of circular barrels, or on opposing linear arcs for angular Ver-

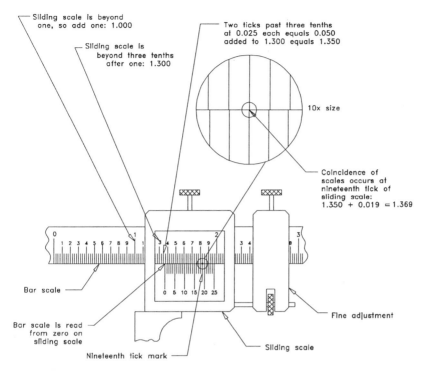

Fig. 2-6. Constituent parts of the English Vernier scale. With practice, this instrument can be used to quickly determine measurements accurate to one one-thousandth of an inch. The Vernier scale method relies on the coincidence of two tick marks on opposing scales.

niers. Linear Vernier scales are typically employed on an instrument called a "caliper," barrel-type Verniers are used on "micrometers," and linear arc Verniers are employed on specialized protractors. Here we will start with an explanation of how to read and use a linear Vernier as a preface to covering the use of the caliper in the next section. This will simplify the explanation of the barrel-type Vernier used on the micrometer when we get to that section. Fig. 2-6 illustrates a typical linear Vernier scale. Note that there are two sets of numbers: one on the "bar," which is a fixed part of the instrument, and the other set on the sliding jaw. The increments, shown here in English units, are divided in tenths, but notice the smallest subdivisions are not fiftieths, rather they are fortieths, or 0.025 of an inch. However, the opposing scale of numbers actually makes this instrument accurate to 0.001 inch.

The incremental tick marks on the sliding scale are spaced such that the 25 graduations cover exactly the same span as 24 graduations of the bar's scale. In other words, each of the graduations on the sliding scale is made 1/25 smaller than the bar scale graduations to allow for one more. If the bar scale graduations are 1/40 of an inch, and the sliding graduations are 1/25 smaller, then 1/25 of 1/40 just happens to equal 1/1000.

Reading the Vernier is done initially from the zero line of the sliding scale. For example, in the illustration, all of the graduations to the left of the zero line of the sliding scale are first added to the distance. First, this means it has passed the one-inch mark; then we count the tenth increments, which in this case are three, so we now have 1.300. Next, each of the smallest tick marks to the left of the sliding scale zero line is added, in this case, there are two ticks after the three-tenths mark still to the left of the zero line. We have already determined that each of these tick marks represents 0.025 inch, therefore, we now add 0.050 to our initial result, i.e., 1.300 + 0.050 = 1.350.

At this point we have exhausted the graduations of the bar scale, now we must account for the graduations on the sliding scale. Each tick mark on the sliding scale represents one one-thousandth of an inch, and is labeled every five-thousandths for ease of reading. The remaining measurement is read by finding the point where *any* tick on the sliding scale coincides exactly with *any* tick mark on the bar scale. In our example, this coincidence occurs at the nineteenth tick mark of the sliding scale. The point of coincidence on the bar scale is immaterial. Since we know that each tick mark on the sliding scale represents one one-thousandth of an inch, all that is left to do is determine the exact point of coincidence and add that tick mark value to the other figure already arrived at. Therefore, 1.350 + 0.019 = 1.369.

Although this may sound like a lot of work—to be constantly adding figures together—it is surprising how quickly an operator will learn to read the main bar scale in graduations of 0.025. It will become automatic to read the bar scale as 1.300, 1.325, 1.350, 1.375, 1.400, etc. Reading the sliding scale will become equally easy with practice, and this will only leave one addition to perform, and that will never be larger than adding 24.

Some English Vernier scales have 50 graduations on the

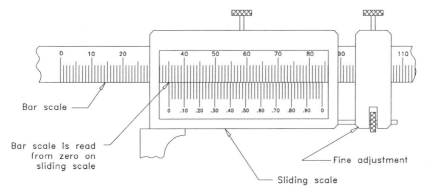

Fig. 2-7. The Metric Vernier scale works very much like the English Vernier scale. This scale can be used to make measurements quickly to an accuracy of 0.02 millimeters.

sliding scale instead of 25. This makes very little difference to the operator. The procedure for reading the instrument is the same: add the measurement from the bar scale to the left of the zero line of the sliding scale to the value of the tick mark of the sliding scale that coincides with any tick on the bar scale. The reason for the difference in these scales is that there are half as many tick mark graduations on the bar scale of these instruments. Therefore, instead of bar scale graduations equaling 0.025 inch, they actually are every 0.050 inch. To compensate for this, the Vernier must span 50 thousandths instead of 25.

The Metric Vernier scale works similarly. Fig. 2-7 illustrates the Metric Vernier scale. The graduations along the bar scale are in millimeters, with numerical values etched at every centimeter (10 millimeters). The tick mark graduations on the sliding scale are placed so that 50 ticks cover exactly the same span as 49 on the bar scale, or, in other words, they are 0.02 mm smaller. Every five ticks is labeled, or every 0.10 mm. Again, the millimeters on the bar scale are read to the left of the zero line on the sliding scale. Then the remaining measurement is counted directly off the Vernier; first, the tenth millimeters, and finally the two one-hundredths of a millimeter are determined by the sliding scale tick mark that exactly coincides with *any* tick mark on the bar scale.

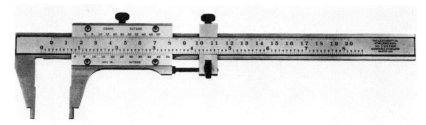

Fig. 2-8. The Vernier caliper can be graduated in both English and Metric units on the same instrument. It can also be designed for both internal and external measurements. In this illustration, the caliper shows Metric measurements along the top and English measurements along the bottom, both for external measurements. However, note the notch in the jaws. This instrument is designed to measure internal dimensions as well, but the operator must remember to add the thickness of the "nibs," which on this particular instrument is 0.250″ or 6.35 mm. (Photo courtesy of The L.S. Starrett Co.)

Calipers

A caliper is an instrument which employs a set of adjustable jaws to contact either the inside or outside of an object. The instrument may be capable of making the measurement directly, or may need to have the span of the jaws measured with a secondary instrument such as a steel rule.

Direct-measurement calipers employ a rule on the bar of the instrument. Depending on the intended accuracy of the instrument, this may or may not be equipped with a Vernier scale. This means that, depending on the type of caliper used, the accuracy may be as low as the capabilities of the associated steel rule, or as high as one one-thousandth of an inch. The direct-measurement caliper is available in a variety of styles, each designed to be slightly easier to read. The Vernier caliper is the most common of the direct-measurement calipers. To aid reading, some calipers employ a dial indicator to the readings in one-thousandths. Finally, in recent years, calipers with electronic digital readouts have been developed. These are by far the easiest to read, but are also the most costly to purchase, and require the greatest amount of delicate handling to maintain their function. Fig. 2-8 illustrates a direct-measurement Vernier caliper with both English System (lower edge) and Metric System (upper edge) capabilities.

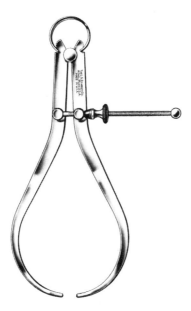

Fig. 2-9. The standard external caliper is in essence an adjustable gap frame. This instrument is only used for taking external measurements. (Photo courtesy of The L.S. Starrett Co.)

EXTERNAL CALIPERS

Measuring an outside dimension and an inside dimension require slightly different techniques. The caliper illustrated in Fig. 2-9 is designed exclusively for taking external dimensions. Whether the part to be measured is flat or round is of no particular consequence, but how the instrument is held relative to the part is very important. Measuring with hand-held instruments relies heavily on the "feel" the operator has for performing the operation. The instrument must be closed on the part completely, but not forced. Additionally, the instrument must be held as square and perpendicular to the part as possible. Learning the correct feel or touch of using a hand-held measuring instrument can only be achieved with practice, patience, and experience.

If possible, the part should be stationary, e.g., fixed in a machine or vice, or at least set on a table or surface plate. This aids in keeping the instrument positioned correctly to the part. If this is not possible, hold the instrument in the dominant hand (right hand if the operator is right handed, left hand if the operator is left handed) and hold the part in the other hand. This is necessary because the operator will need the added dexterity of the dominant hand to operate the jaws of the caliper (and the fine adjustment, when applicable).

The sliding jaw of the caliper is moved toward the part until it contacts it. Most calipers are equipped with a fine adjustment if they lack dial or electronic capabilities. Once the sliding jaw and fixed jaw contact the part, tighten the screw on the top of the fine adjustment. Only then begin moving the sliding jaw toward the part with the fine adjustment screw while simultaneously making sure to align the instrument so that the flat of the jaws is completely contacting the part squarely. Once the jaw will no longer move forward with modest pressure from the adjusting screw, stop and take the reading. Never force a measurement instrument! If necessary, clamp the sliding jaw in place with its clamping screw to assure that the caliper remains in position while taking the reading. This eliminates the possibility of twisting the instrument slightly while removing it from the part and consequently forcing the jaws apart slightly resulting in an inaccurate measurement.

INTERNAL CALIPERS

The internal caliper works very much like the external caliper with a few differences. The indirect-measurement internal caliper is a reverse of the external caliper, with the contact measuring points turned out instead of in. Fig. 2-10 illustrates an indirect internal caliper. The direct-measurement Vernier caliper employs a completely different type of jaw from its external measuring cousin.

Fig. 2-11 illustrates a combination internal–external dial Vernier caliper. This particular type of Vernier caliper does not use a fine adjustment feature, so it is necessary to clamp the sliding jaw directly while holding the caliper in position. The dial indicator replaces the Vernier scale for reading the finest graduation of one-thousandth inch. Note that the internal caliper jaws (on the top of the bar in this particular caliper) are actually "blades" that pass each other to arrive at a distance of zero.

Each of these blades has a near knife edge where the instrument will contact the part for internal dimensions. This feature

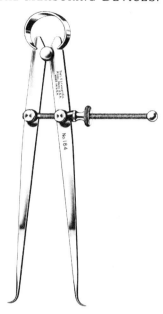

Fig. 2-10. The internal indirect caliper, like its external cousin, is only capable of internal measurements, and must be used with an associated steel rule for determining the measurement. This instrument would only be used for measurements with a comparatively low tolerance due to possible reading errors and lack of a Vernier scale. (Photo courtesy of The L.S. Starrett Co.)

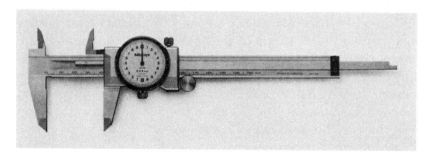

Fig. 2-11. The Vernier caliper can be augmented with a dial for reading the smallest part of the fractional unit. This particular caliper is a Metric instrument for both internal and external measurements with an accuracy of 0.02 mm. The dial caliper is read much more quickly than the standard Vernier caliper, but is more costly and slightly more delicate. However, a careful operator will get many years of service from this instrument and save a great deal of time in making measurements. (Photo courtesy of Mitutoyo/MTI Corp.)

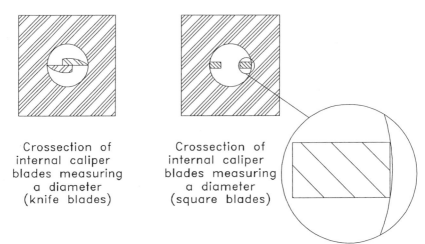

Crossection of
internal caliper
blades measuring
a diameter
(knife blades)

Crossection of
internal caliper
blades measuring
a diameter
(square blades)

Fig. 2-12. Cross section, looking down, of a hole measured with an internal caliper, one with knife edge jaws and one with square jaws. The resultant error is obvious. If the internal caliper jaws were made square, the instrument would actually be sitting on a tangent point creating a crown height error under the end of the jaw. This error would of course be doubled because it would occur at both jaws.

is used for internal calipers for a very logical reason. For example, consider measuring between two flat surfaces for an internal dimension. In this case, the shape of the caliper jaw would be immaterial. Now consider measuring the diameter of a hole with the internal caliper. If the jaws were anything other than a knife edge or point, the width of the jaw blade would contact the internal diameter at a tangent point and give an inaccurate measurement. Fig. 2-12 is a cross section looking down into a hole measured with an internal caliper—one with knife edge jaws and one with square jaws. The resultant error is obvious.

Knife edge internal caliper jaws alone will not assure accurate measurement of an internal surface. Again, we come back to the fact that a great deal of the use of hand-held measuration instruments relies on the ability of the operator to "feel" the correct positioning of the instrument. For example, if the operator is measuring between two flat internal surfaces, then it is imperative that the instrument be held exactly perpendicular to these surfaces to avoid measurements which are too large. By moving the instrument back and forth slightly while applying minimal pressure outward, the operator will be able to feel where the shortest distance is being measured. Conversely, if the operator is measuring the diameter of a hole, then it is necessary to make sure that the jaws contact the hole at the greatest possible distance to avoid taking a measurement that is too short.

Given all of this necessity for human finesse when taking internal measurements with the internal caliper, it may seem this is not the ideal instrument for making these sorts of measurements. This may or may not be true, depending on two factors: the time allotted for taking the measurement, and the degree of accuracy necessary from the measurement. The internal caliper is a relatively accurate instrument, down to one one-thousandth of an inch or two one-hundredths of a millimeter, and can be employed with great speed by an experienced operator with a developed touch. However, if greater accuracy is required, and repeated checking of an identical hole is required, another instrument is probably a better choice.

There are specialized instruments used for taking internal diametrical measurements, and these vary greatly depending on factors such as available time, degree of required accuracy, and number of repetitions. Internal micrometers are more accurate, and custom gauges are as accurate (or more so) and can be employed with greater speed if the number of holes to be measured is sufficient to justify making a gauge. We will cover these other internal measuring devices later in the book.

DEPTH GAUGE

The caliper is sometimes considered a modification of the steel rule by adding jaws and a Vernier scale. The depth gauge is probably even closer to the steel rule than the caliper. This instrument consists of a bar nearly identical to a steel rule and a Vernier scale. The Vernier scale is slightly modified with a base to rest the depth gauge on the surface of the part being measured. Fig. 2-13 illustrates a Vernier depth gauge. A similar instrument is the depth gauge with no Vernier scale attachment. Measurements are then made directly as they would be with a steel rule.

The depth gauge is placed on the part in such a manner so that one or both of the flanges of the Vernier base are setting completely flat on the part. The rule, or bar, is then extended

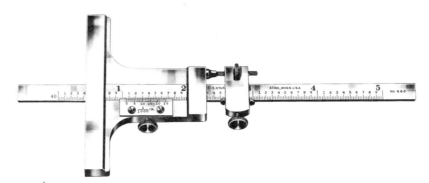

Fig. 2-13. The Vernier depth gauge is in essence an open frame micrometer with a cylindrical scale. Note that this particular instrument is supplied with interchangeable spindles for progressively deeper recesses. (Photo courtesy of The L.S. Starrett Co.)

down to contact a lower surface. Fine adjustment is accomplished as with a caliper—the instrument is held in place with one hand while the operator clamps the fine adjustment screw and then adjusts the bar up or down so that the Vernier head and rule end are both contacting their appropriate surfaces. The Vernier head is then clamped to maintain correct reading, and the instrument is extracted from position and the measurement read. These instruments are also available with digital electronic readouts.

GEAR TOOTH CALIPER

The gear tooth caliper is exactly what the name implies: a specialized caliper used exclusively for measuring the cord thickness of gear teeth at the tooth's pitch diameter. Fig. 2-14 illustrates measuring a gear tooth. (See Appendix C for details.) Taken in this light, the gear tooth caliper acts as both a Vernier depth gauge and a Vernier caliper. The vertical slide acts as the depth gauge and the horizontal slide acts as the Vernier caliper. The Vernier scales on each of these slides are read like any other Vernier scale.

This instrument is predominantly a checking instrument. First, the depth of the pitch diameter is set on the vertical slide and the slide is locked in place with the clamping screw. The

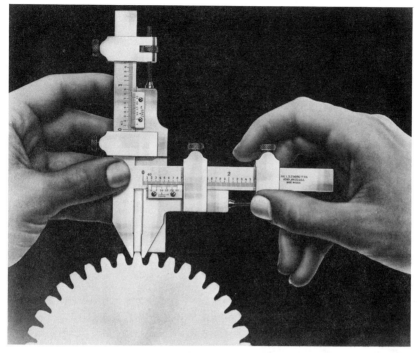

Fig. 2-14. The gear tooth Vernier caliper acts as both a depth caliper and a width (external) caliper. The instrument measures the cord width of a gear tooth along the pitch diameter of the gear. The depth portion of this instrument allows the operator to set the jaws at the proper depth for measuring at the gear's pitch diameter. (Photo courtesy of The L.S. Starrett Co.)

instrument is then placed down over the gear tooth, and the horizontal slide caliper is closed and adjusted over the cord of the gear tooth. The caliper can then be locked in place and removed to take the measurement reading.

Micrometers

The micrometer is similar in function to the Vernier caliper, but bears little resemblance to one, and almost none to the standard non-Vernier caliper. While the micrometer operates like the caliper, it employs a precision screw thread to actuate the movable "jaw." The "jaws" of a micrometer are called the "spindle" (for the movable one) and the "anvil" (for the fixed one). The

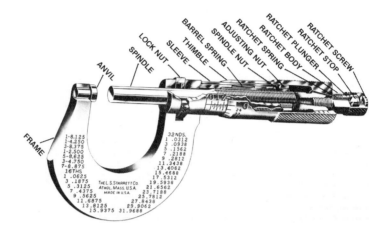

Fig. 2-15. The micrometer is a highly intricate and delicate instrument. With sufficient care, however, it will give years of accurate service. (Photo courtesy of The L.S. Starrett Co.)

structure of the micrometer varies radically depending on the intended application of the instrument, and the shape of the anvils will also vary radically even among similar micrometers to accommodate different measuring applications. Fig. 2-15 is a cutaway illustration of a standard external micrometer.

The micrometer is typically made with 40 threads per inch on the actuator screw, so that one full revolution of the handle will advance the spindle 1/40 (0.025) of an inch. One of the main differences between the caliper and the micrometer is that the latter *always* employs some sort of a Vernier scale, even if that scale is digital or electronic digital. The micrometer is a more precise instrument than the caliper. The Vernier scale of the micrometer is also somewhat different than that of a linear caliper, and this is complicated by two degrees of accuracy added to the two measurement systems. Therefore, before we examine the different types of micrometers, we will detail how to use and read this type of Vernier scale.

THE CYLINDRICAL VERNIER SCALE

We mentioned that the actuator screw of the micrometer spindle is a precision thread with 40 threads per inch so that each complete revolution of the "thimble" advances the spindle

1/40 (0.025) of an inch. If you think back to our discussion of the linear Vernier, this is also the same amount of each graduation on the bar scale of the caliper. However, this thread per inch ratio was not chosen by chance or tradition. It was in fact the only way to make the Vernier scale the same between instruments. Therefore, learning one automatically lends itself to learning the other with ease.

Micrometers Calibrated in Thousandths: Fig. 2-16 illustrates the cylindrical Vernier scale calibrated in thousandths of an inch, with the cylindrical scale of the thimble laid out flat. The graduations which run laterally down the "sleeve" of the scale represent 1/40 of an inch. In other words, each tick mark that is exposed from under the "thimble" of the micrometer equals 0.025 inch which the micrometer's anvil gap is open. Every fourth tick mark is etched with a number indicating the number of tenths (0.100) it represents.

The Vernier scale, which is movable and necessary to make measurements of one-thousandth of an inch, is etched around the circumference of the thimble. There are 25 tick marks around the circumference of the thimble, exactly corresponding to the 25 tick marks we encountered on the linear Vernier scale. This is why it is vital that the actuator screw have exactly 40 threads per inch. To facilitate alignment of the thimble's Vernier tick marks,

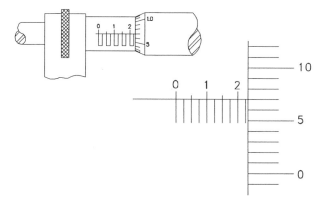

Fig. 2-16. The cylindrical Vernier scale of a micrometer calibrated in one-thousandths of an inch. The reading shown in the enlargement of this cylindrical Vernier would be: 0.1 + 0.025 + 0.007 = 0.232.

0.2

a line is etched transversely across the tick marks of the sleeve. When the spindle is brought to bear against a part to be measured, first the increments exposed on the sleeve are determined, then the tick mark on the thimble which comes closest to alignment with the transverse line of the sleeve is added.

Micrometers Calibrated in Ten-Thousandths: Now let's examine the next most precise instrument in the linear measurement family—the micrometer calibrated in ten-thousandths. This instrument looks very much like the micrometer calibrated in one-thousandths of an inch, with one very subtle difference—the addition of a second Vernier scale on the sleeve. Fig. 2-17

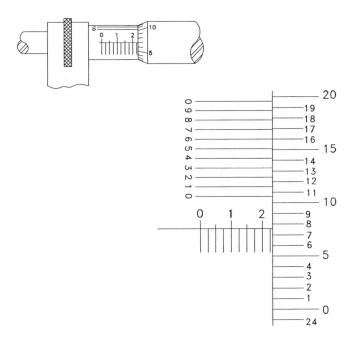

Fig. 2-17. The cylindrical Vernier scale calibrated in one ten-thousandths of an inch is very similar to the same device calibrated in one-thousandths, with the addition of a second Vernier scale attached to the first scale. In the enlargement in this illustration, the reading would be 0.2 + 0.025 (both from the lateral sleeve scale) + 0.007 (from the thimble scale) + 0.0006 (from the cylindrical sleeve scale) = 0.2326. The number 7 is used from the thimble scale even though it is past the reading line because it is less than 8, and we know the remaining amount will come from the last scale on the sleeve.

illustrates the Vernier scales used with the micrometer calibrated in ten-thousandths of an inch. This second Vernier scale is placed along the circumference of the sleeve just above the lateral number scale. This additional Vernier scale consists of ten tick marks which cover exactly the span of nine tick marks on the thimble. This makes each of the sleeve ticks one-tenth smaller than the thimble ticks, the difference therefore being one-tenth of one-thousandth, or one ten-thousandth.

Reading the ten-thousandths on this Vernier is accomplished very much like reading the one-thousandth marks on a linear Vernier. Once the spindle comes to rest, first the values transversing the sleeve are determined, then the value of the tick mark on the thimble is determined, just as with a micrometer calibrated in one-thousandths. Finally, the last Vernier scale for ten-thousandths is examined. The tick mark on the sleeve coinciding with any tick mark on the thimble is the value of the ten-thousandths. The three values are added together to obtain a result.

Micrometers Calibrated in Hundredths of a Millimeter: The metric micrometer obviously will have modifications in the structure as well as the etched tick marks to compensate for the difference in measurement systems. For example, the actuator screw employs a metric thread with a thread pitch equal to one-half millimeter per thread. This means that one complete revolution of the thimble will advance the spindle one-half millimeter. This is done for the same reason that 40 threads per inch was chosen for the English System micrometer: the thread pitch is directly related to the Vernier scale.

The metric micrometer calibrated in one-hundredths of a millimeter looks and works very much like the English one-thousandths micrometer. The main scale is etched laterally along the length of the sleeve. The difference is that this scale has whole units above the transversal line, and the fractional half millimeters etched below this line. Every fifth millimeter is numbered over the length of the capacity of the instrument.

The Vernier scale is etched around the circumference of the thimble, with a total of 50 tick marks. As might be deduced, each of these tick marks represents 1/50 of the smallest graduation on the sleeve or, in other words, 1/50 of one-half millimeter,

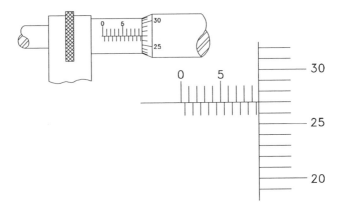

Fig. 2-18. The Metric cylindrical Vernier scale calibrated to one-hundredths of a millimeter is slightly different than its English one-thousandth cousin. The lateral scale of the sleeve is divided along the top of the reading line at every millimeter; and along the bottom of the reading line are intermittent tick marks representing half-millimeters (0.5 mm). The cylindrical scale of the thimble has tick marks representing the one-hundredth millimeters (0.01 mm). Therefore, the enlarged flat pattern of the cylindrical scale in the illustration would read 9.0 + 0.5 + 0.27 = 9.77 mm.

which is one one-hundredth of a millimeter. Fig. 2-18 illustrates the Vernier scales used for determining measurements of metric dimensions in hundredths of a millimeter.

Micrometers Calibrated in Two-Thousandths of a Millimeter: The metric micrometer calibrated in two-thousandths of a millimeter is very similar to the English micrometer calibrated in ten-thousandths of an inch. The two-thousandth (0.002) measurement is effected by the addition of a second Vernier scale along the circumference of the sleeve. This Vernier scale consists of five tick marks, each of these representing one-fifth of a thimble graduation. Therefore, one-fifth of one-hundredth is two-thousandths. The tick marks are each etched with a numerical value incrementing by two. Fig. 2-19 illustrates this metric Vernier scale. The two thousandths increments are read directly from the Vernier scale on the sleeve by their coincidence with any tick mark on the thimble.

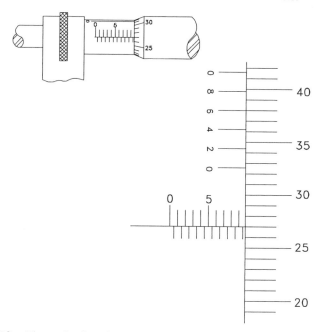

Fig. 2-19. The cylindrical Metric Vernier calibrated to two one-thousandths of a millimeter adds a second Vernier scale to the sleeve to read
the one-thousandths of a millimeter. This scale, like all of the other
Vernier scales, is read from the point of coincidence of any tick mark on
the two opposing scales. In this particular illustration, the instrument
would read 9.0 + 0.5 + 0.27 + 0.006 = 9.776 mm.

EXTERNAL MICROMETERS

The external micrometer looks vaguely like a C-clamp, in that
it has a solid "C"-shaped frame which the spindle travels across
to close the gap of the frame. The spindle and anvil are precision
ground to be completely flat and parallel to each other. Fig. 2-20
is an external micrometer with a standard Vernier scale. New
technologies are being added to the micrometer to make reading
the instrument easier and faster for the operator. Fig. 2-21 is a
micrometer with the addition of an electronic digital readout,
automatic switching between English units and Metric units, and
electronic zeroing of the instrument.

The micrometer is typically designed with a comparatively
short range of possible measurements. Usually the travel on the
spindle is only an inch or two. Obviously this means that the size

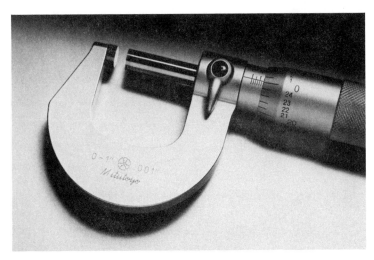

Fig. 2-20. A typical micrometer for taking very accurate measurements of small parts. The spindle and anvil of this instrument are ground and polished to a high degree of smoothness and to be exactly parallel surfaces. Note the locking lever on the side of the frame for fixing the spindle after taking a measurement. (Photo courtesy of Mitutoyo/MTI Corp.)

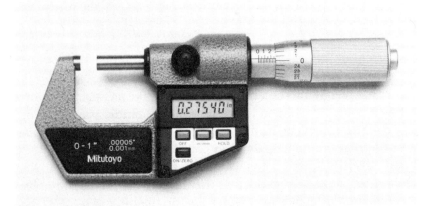

Fig. 2-21. The external Vernier micrometer is available with either mechanical digital readout or, as shown here, with electronic digital readout. Digital readout micrometers are easier to read and faster to operate. They are also typically more expensive to purchase and slightly more delicate, although with proper handling they will give the operator just as many years of service as a standard Vernier micrometer. (Photo courtesy of Mitutoyo/MTI Corp.)

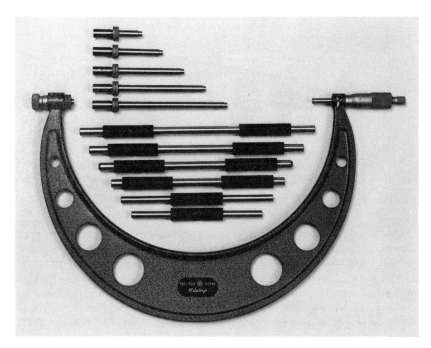

Fig. 2-22. The large-frame Vernier micrometer is furnished with a set of progressively longer anvils to accommodate various measurements with a limited travel spindle. Note the rods, which are not anvils, located within the frame of the micrometer that have black bands on them. These are "standards," which are metal rods finished to exact lengths for calibrating the instrument after changing anvils. The black bands are insulated handling strips to keep the standard from absorbing heat from the operator's hand and thus expanding. Expansion of steel due to change in temperature is equal to approximately 0.000006 inch per linear inch of material for each 1°F rise in temperature. Therefore, a 6-inch standard raised 7° in temperature would "grow" about 0.000252 inch (0.000006 × 6 inches × 7°). (Photo courtesy Mitutoyo/MTI Corp.)

of the micrometer must be fairly close to the size of the object being measured. Therefore, to accommodate measuring larger objects, a micrometer with a larger frame must be used. Fig. 2-22 illustrates a large-frame micrometer. Note that this instrument maintains a set of interchangeable anvils to accommodate various size measurements. The anvil is changed to bring the gap between the anvil and spindle down to less than one inch, because the travel of the spindle is only one inch maximum.

Fig. 2-23. The sheet metal micrometer is a type of standard Vernier micrometer with a specially designed frame to allow access to measure parts at points far from the edge which are comparatively thin. Note that this instrument, like most other micrometers, has a fairly limited amount of travel for the spindle. This instrument is also provided with a tripod mount to rest it on a workbench, due to the difficult nature of holding the instrument straight by hand. (Photo courtesy Mitutoyo/MTI Corp.)

The other potential situation is that the part being measured is thin but wide, as is typical with sheet metal parts. To accommodate this, the frame of the micrometer is designed with a small gap, but with a very deep frame. Fig. 2-23 illustrates this type of micrometer used for wide, thin parts.

INTERNAL MICROMETERS

The internal micrometer, in the strictest sense, is a micrometer with the spindle and anvil in line on opposite sides of the thimble. This device is used to make a two-point contact at the "shortest" possible distance between two opposing internal surfaces. Fig. 2-24 is a standard internal Vernier micrometer.

This particular instrument will be used to determine the "shortest" possible distance between two internal surfaces, provided those surfaces are flat. If the surfaces are opposing sides of a diametrical hole, then it is used to measure the "greatest" possible distance. This will obviously require a developed "touch" by the operator to make sure the largest possible distance is being measured.

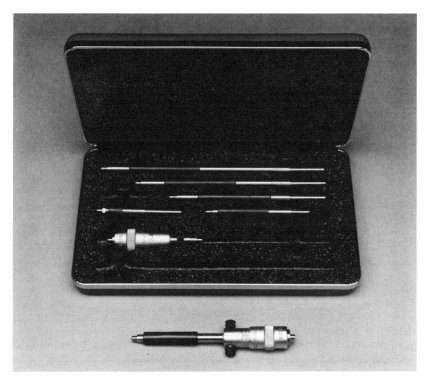

Fig. 2-24. The standard internal Vernier micrometer is designed with the anvil, spindle, and thimble all in line. Multiple interchangeable anvils allow this instrument to be used over a wide range of measuring applications. The Vernier scale is read in exactly the same manner as any other cylindrical Vernier scale. (Photo courtesy The L.S. Starrett Co.)

However, when the internal surface being measured is diametrical, and small, the internal micrometer of this type is difficult or even impossible to use. To make the measuring of these internal diameters easier and more accurate, there is yet another type of internal micrometer. This instrument uses three spindles which extend out from a central shaft. The three-point contact automatically centers the instrument in the hole. The calibration of the instrument is designed such that it will interpret, for the operator, the diameter of the hole.

Fig. 2-25 is a similar instrument called a "bore gauge." This instrument works essentially like the three-point internal mi-

Fig. 2-25. The bore gauge is a type of specialized internal micrometer for measuring internal diameters using a three-point contact. The instrument is designed to automatically calculate the diameter of the hole, and the resultant measurement is displayed on the dial indicator near the handle. (Photo courtesy of Federal Products Co.)

crometer, but in lieu of a standard Vernier scale on the thimble, it employs a dial indicator to read the measurement. We will discuss the construction and use of the dial indicator later in this book. The dial indicator on this instrument displays the measurement of the bore gauge in the same way the dial of the "dial Vernier caliper" displays the readings of that instrument. The bore gauge used for measuring shallow holes with a two-point contact is illustrated in Fig. 2-26.

Depth Micrometer

The depth micrometer is very similar to the depth gauge (caliper) except that it is much more accurate. The instrument has a "T"-shaped frame with the spindle extending from the center of the frame. The depth micrometer is rested on one surface of the object being measured and the spindle is extended down to a lower surface, such as the bottom of a bored hole.

Fig. 2-27 illustrates the depth micrometer. Note that this

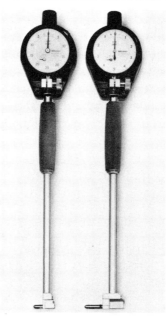

Fig. 2-26. Bore gauges for shallow holes are essentially internal micrometers which employ a two-point contact system. The measurement is registered on the dial indicator at the end of the handle. (Photo courtesy of Mitutoyo/MTI Corp.)

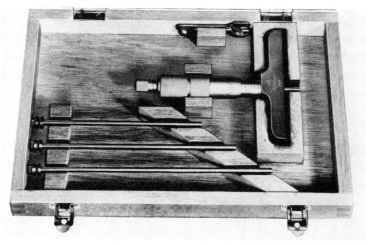

Fig. 2-27. The Vernier depth micrometer. This instrument measures the distance between two different levels, or surfaces, on a part. The anvil is missing on this instrument because the part surface actually replaces the anvil. The depth micrometer is supplied with a set of progressively longer spindles to accommodate deeper measurements with a limited travel of the spindle. (Photo courtesy of Mitutoyo/MTI Corp.)

instrument is supplied with a series of progressively longer "spindles." Like the large-frame external micrometer discussed earlier, the spindle travel of this instrument is limited to a comparatively short range, typically one inch. However, unlike the large-frame micrometer, this instrument has no "anvil." Therefore, to accommodate holes with a depth beyond the range of the spindle travel, the spindle is interchangeable with a longer spindle. This allows for one instrument to be used in a wide variety of measuring applications without the need for additional depth micrometers. In the case of this instrument, the lower surface of the object being measured acts as the "anvil" of the instrument.

MISCELLANEOUS MICROMETERS

The three micrometers discussed so far are essentially the only types of micrometers that exist. The addition of digital or electronic digital readouts only make the instrument more "user friendly," it doesn't change the basic function of the instrument. Likewise, there are various spindle and anvil configurations which are available which will give the micrometer a specialty name, but will not significantly change the basic function or reading of the instrument.

Fig. 2-28, for example, is a "point micrometer." This instrument uses the basic external micrometer frame, but has a pointed anvil and spindle. The point micrometer is used for measuring

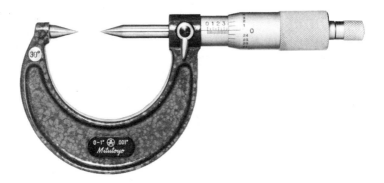

Fig. 2-28. The "point micrometer" is essentially a standard external Vernier micrometer, with specialized anvil and spindle for measuring hard-to-reach places. Aside from the reach of the instrument, it is used and read exactly like the regular external Vernier micrometer. (Photo courtesy of Mitutoyo/MTI Corp.)

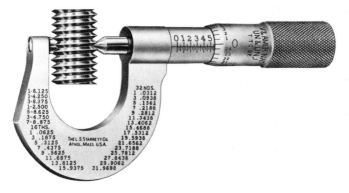

Fig. 2-29. The screw thread is another specialty form of the external micrometer. The spindle and anvil are specially designed to measure threads at the pitch diameter. The remainder of the instrument itself is used in the same manner as any other external micrometer. (Photo courtesy of The L.S. Starrett Co.)

inside hard-to-reach places, e.g., measuring to the bottom of a groove. Fig. 2-29 uses a standard external micrometer frame, but with a specially shaped anvil and spindle to reach the "pitch diameter" of a screw thread. This instrument is called a "screw thread micrometer" even though only the anvil and spindle have been replaced.

Finally, Fig. 2-30 is a specialty micrometer that actually does have a somewhat different configuration in the frame. This instrument is called a "Uni-mic." The Uni-mic has a frame in which the anvil is fastened to the outside of the frame. The anvil can then be easily replaced with multiple, different configurations to accommodate any of a variety of different measuring applications. In this illustration, however, the anvil has been completely removed and the instrument is being used in conjunction with a "surface plate" to operate like a miniature "height gauge." The measurement is taken between the resting place of the instrument and the spindle. We will discuss surface plates and height gauges later in the book.

Dial Indicator

The dial indicator has been mentioned briefly as an add-on feature of the Vernier caliper, but it is also used as an indepen-

Fig. 2-30. The multiple anvil, or universal anvil, micrometer employs any of a number of different anvils with a standard flat spindle for a variety of measuring applications. Shown here with the anvil removed entirely, the instrument is used in a similar manner to a miniature height gauge with a small surface plate. (Photo courtesy of Mitutoyo/MTI Corp.)

dent measurement instrument. By independent, we mean it is the actual device doing the measurement, but it is invariably attached to some other device because it has no jaws like a caliper, nor a stand of its own like other instruments. The stand can be a specially designed dial indicator stand, either with or without a magnetic base, for use on a machine table or a surface plate. The dial indicator can also be attached to a height gauge, a machine spindle (like that of a milling machine), or a tool holder (like that found on a lathe). Fig. 2-31 illustrates a dial indicator on a magnetic base stand for use on a machine table.

Fig. 2-32 illustrates the face and the internal mechanism of the dial indicator. The instrument is designed with a spindle that travels up and down through a bushing, and the spindle has a rack gear cut into the side. This rack interfaces with a round gear, which is part of a very accurate gear chain that in turn operates the needle indicator on the face of the dial. The end of the spindle is capped with a hardened steel tip to reduce wear of the spindle;

Fig. 2-31. The dial indicator mounted on a special dial indicator stand with a magnetic base for use on a machine table. (Photo courtesy of Federal Products Co.)

this is because the dial indicator in this form is a full contact measuring device that must slide across the surface of the part it is measuring.

The purpose of the dial indicator is to check a part's surface contour or position. For example, to check the runout of a surface, the part is placed with one face on a reference plane, such as a mill table or surface plate, then the dial indicator attached to a stand is brought into contact with the top surface of the part until the needle registers at the "bottom" of the dial. Next, the needle is adjusted back to zero by means of the small adjusting knob on the side. Finally, the stand of the dial indicator is moved along the part to allow the dial indicator to take readings along the entire surface.

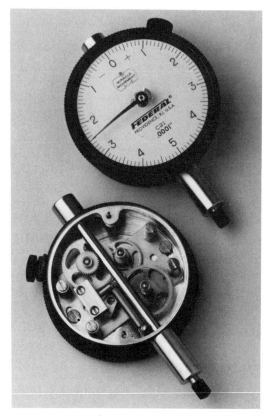

Fig. 2-32. The face and internal mechanism of the dial indicator. (Photo courtesy of Federal Products Co.)

The indicator is forced down so the needle is at the bottom, and then zeroed, because this allows the instrument to read values that are either positive or negative from the point of beginning and to assure correct contact. Note in the illustration that there is a plus and minus sign on either side of the zero on the dial, and that the numbers progress from there in both directions. As the end of the spindle travels along the workpiece, if the surface is running down toward the reference plane, the needle will show a negative value; and, conversely, if it is running away from the reference plane, it will indicate a positive value.

Applications of this instrument are nearly limitless. Placed in a lathe tool holder, the dial indicator can be used to check roundness or concentricity. Mounted in the spindle of a milling ma-

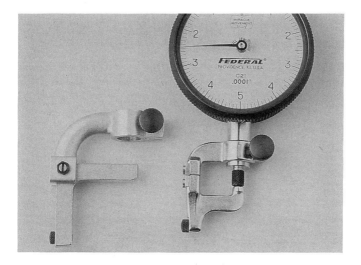

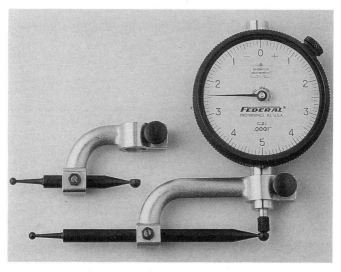

Fig. 2-33. Special attachment heads for taking measurements in close spaces or at right angles with a dial indicator. (Photo courtesy of Federal Products Co.)

chine, it can be used to test for runout of the machine's table. If the part is to be checked on a surface plate, but the bottom of the part is not parallel to the surface facing the surface plate, the part can be attached to another device and the dial indicator used to set up the top surface so it is parallel to the surface plate.

With the addition of special spindle attachments, the dial

indicator can be used to check surfaces in nearly any attitude. This is especially important when checking areas that would not be accessible with the standard dial indicator spindle, such as inside a slot or an internal diameter. Fig. 2-33 illustrates some of the many spindle attachments available for use with the dial indicator.

Chapter 3

Squares and Surface Plate Instruments

DESPITE MANKIND'S FASCINATION with a multiple of geometric forms, in industry most things need to be "square." In this case, "square" does not refer to the geometric shape, but rather to the relative position of two opposing planes, or surfaces, of a workpiece. To be "square" to each other, two lines or planes must form an angle of 90° to one another. The angle formed between two planes (or surfaces, as is typical in industry) is called a "dihedral angle."

The reason for the importance of squareness is the manner in which most parts are measured and manufactured. Typically, three-dimensional objects are measured in space by relative distances between two points, but that distance is rarely a straight line. For example, in Fig. 3-1, the machinist must locate the center of a hole to be drilled in a plate of steel. Notice that the location of the hole is not on a straight line between the point of origin and the center, but is a series of relative distances from the origin that are either parallel or perpendicular to the earth.

The other option would use a straight line between the point of origin and the hole center, and it would have to include not only the straight line distance, but the angle from one of the edges of the workpiece to the line formed between the origin and the

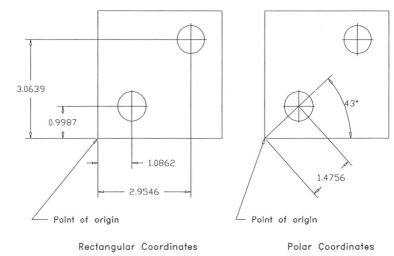

3.0639

0.9987

1.0862

2.9546

Point of origin

43°

1.4756

Point of origin

Rectangular Coordinates   Polar Coordinates

Fig. 3-1. Illustration of the differences between standard rectangular (Cartesian) coordinates and polar coordinates.

hole center. This method, called "Polar Coordinates" or "Vector Coordinates" is a viable method, but is more difficult for the machinist to apply in most applications.

The method of using straight line distances parallel or perpendicular to the earth is called "Cartesian Coordinates." The advantage to this system is that every distance is measured in a direction that is square to every other distance. This system is simply a lot easier to use and largely explains why squareness is so important to the machinist. We will cover the Cartesian Coordinate System in depth later in the book under the section on Coordinate Measuring Systems.

When taking measurements like those in our illustration, squareness is critical. If the measurements were taken in this system with distances that were out of square, the locations would be inaccurate. This is especially true as the distances become longer, because the measurement is actually being taken along the hypotenuse of a triangle.

The Surface Plate

As might be expected, measurements involving squareness must originate from some place that can be used as a bench-

Fig. 3-2. A typical surface plate for taking precision measurements. This particular surface plate is made of granite, although some surface plates are made of steel. (Photo courtesy of The L.S. Starrett Co.)

mark. This plane of origination must be absolutely flat, relatively parallel to the earth, and not subject to distortions caused by temperature variations, humidity changes, pressure changes, or any other environmental changes. The "surface plate" is the instrument used for this plane of reference. It may be difficult at first to think of the surface plate as a delicate instrument, since it typically consists of either a solid steel plate several inches thick or, more commonly, a slab of polished granite several inches thick. Yet despite this instrument's substantial construction, it is a highly accurate and, in certain respects, very delicate instrument.

Fig. 3-2 is a typical industrial surface plate made from polished granite. While this is a comparatively hard surfaced material, it is not completely impervious to scratches, gouges, impact dents and chips, or every known solvent. It will withstand a lot of wear, but not much abuse. Like any instrument, the surface plate is available in a variety of qualities, and selection is based on intended end use and cost. Quality of the surface plate is divided into categories: wear resistance and flatness of the surface finish.

The wear resistance of the surface plate is determined by the material from which it is made. Granite is substantially harder than steel, but will also vary in hardness depending on its content of mica and crystalline quartz. This material composition will also determine the surface plate's resistance to changes in temperature.

The surface finish of the surface plate is determined by the manufacturer, and is graded according to the intended end use. Surface plates fall into one of three grades: laboratory grade, inspection grade, or toolroom grade. Laboratory grade surface plates can have surface flatness with as little as 0.000035 (35 millionths) inch deviation in flatness, and toolroom grade surface plates as little as 0.000110 (110 millionths) inch deviation in flatness. Even very large toolroom grade surface plates will have a surface finish with no more than 0.0015 inch deviation in flatness.

Care should always be taken to see that the surface plate is kept clean with an appropriate surface plate cleaning solution to avoid having metal or grinding fragments (which, as might be expected, are very hard and very sharp) rub against the surface plate when in use. When not in use, the surface plate should be covered to avoid possible impact and to protect it from objects that might be set on it. Care should always be taken when placing workpieces and other instruments on the surface plate to reduce wear and avoid damage to the surface finish.

Measurements taken on a surface plate typically involve placing the workpiece on the surface and using another instrument such as a height gauge, steel rule, multiple anvil micrometer, or the like to measure from the surface plate to a given point on the workpiece. If the workpiece is odd shaped or must be in an awkward position to take the measurement, then it will be attached to some other device that sets on the surface plate to assure accurate measurement. Specially made angle brackets, risers, V-blocks, and other devices are made to exacting tolerances for use with the surface plate. Many of these are available in either steel or granite. Fig. 3-3 illustrates some of the common granite surface plate accessories used for positioning the workpiece parallel to the surface plate.

When taking measurements of a workpiece on a surface plate, it is absolutely necessary to be certain that the surface being measured is perfectly perpendicular (square) to the top of the surface plate. This is usually the first consideration when using a surface plate, and can be determined with reasonable speed and accuracy with a precision square. In the next section we will discuss precision squares; although not all of them are used in

Fig. 3-3. Auxiliary surface plate instruments made from granite are a very stable and accurate means of positioning a workpiece parallel to the surface plate. (Photos courtesy of The DoALL Co., Des Plaines, IL.)

conjunction with a surface plate, many can be and a few absolutely must be.

Precision Squares

The square is a fairly simple and yet vital instrument. In one form or another, it consists of two surfaces which are positioned exactly 90° from one another. Usually these two surfaces take the form of blades, but there are certain exceptions to this rule which we will cover in this section. In blade-type squares, either the inside edge or the outside edge can be used to check perpendicularity. This is advantageous because it allows one instrument to be used for checking both inside and outside surfaces of a workpiece.

THE T-SQUARE

Fig. 3-4 illustrates a typical precision square. This particular instrument is called a "Tri-Square" or simply "T-Square." The wider of the two blades can be used in one of several ways. For determining straight line distances away from a surface, the wide blade is allowed to hang over the edge and butt against it. The remaining blade will therefore sit flat against the surface where the measurement is being taken. Frequently, the narrow blade of the T-square is etched with a measurement scale similar to the steel rule to facilitate rapid measurement.

This instrument can also be used in conjunction with the surface plate. The wide blade is rested on the surface plate thus allowing the narrow blade to form a perfect right angle with the

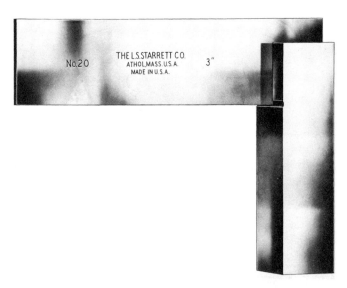

Fig. 3-4. Typical industrial precision square with fixed blades. Sometimes this instrument is called a tri-square or simply a T-square. (Photo courtesy of The L.S. Starrett Co.)

plate. In this fashion, the T-square can be used to determine if (or set up) a workpiece perfectly square to the surface plate. This is done prior to taking any measurement on the workpiece from the surface plate to ensure accuracy of the measurement. If only fairly rough measurements are required, the rule etched on the square's vertical blade can be used for taking the measurement. If a more precise measurement is required, the square is used first to obtain perfect squareness to the surface plate, and then another instrument such as a height gauge is used for taking the linear measurement.

Since this instrument is frequently used for layout work as well as inspection, it is necessary to be able to scribe a line with a hardened scriber point along the edge of the blade. The blades of these squares are hardened to resist wear from this type of layout work. One variation of this instrument is sometimes employed to facilitate easier and more accurate layout. The edge of the narrow blade is beveled to reduce the thickness. Although it is never a knife edge, it is very thin directly at the edge of the blade, allowing only the side of the scriber point to contact the blade. This specialized form of the T-square is sometimes called a "bevel-edge square."

The Cylindrical Square

The cylindrical square is the exception to many of the rules about precision squares. Physically it looks more like a soda pop can than a precision measuring instrument. Fig. 3-5 is an illustration of a cylindrical square. This instrument has no blades, as such, and consists of a single heavy-walled tube that is ground and polished to near-perfect "cylindricality." The ends of the square are then ground and lapped to be completely perpendicular to the axis of the cylinder.

The cylindrical square is typically used as a master square to check the accuracy of other squares because of its exceptionally accurate standards of manufacture, and because the very nature of the instrument will only allow for a single line of contact between the square and the item being checked. Some cylindrical squares are manufactured with four grooves cut on the ends to facilitate the collection of dust particles. This instrument is twisted slightly to force the dust particles on the surface plate

Fig. 3-5. The cylindrical square looks more like a soda pop can than a typical square, but it is in fact the "master square" used for checking the accuracy of other squares. (Photo courtesy of The DoALL Co., Des Plaines, IL.)

into the groove and away from the contact surface. Obviously, this particular square is one that *must* be used in conjunction with a surface plate to function.

One variation of the cylindrical square is the "direct reading cylindrical square." This instrument has one end ground and lapped perfectly square to the axis of the cylinder, and the other end ground slightly out of square. On the side of the instrument is etched a series of curving lines of dots, each terminating in a number. The last row of dots is a straight vertical line terminating in the number zero. The instrument is placed on the surface plate with the out-of-square end on the plate, and placed up against the part to be checked. As the direct reading cylindrical square is rotated against the workpiece, it will eventually form a complete line of contact with the part. The uppermost line in contact with the part will terminate in a number indicating the amount that the workpiece is out of square.

The cylindrical square, like all instruments, needs to be handled with care, even though it is basically only a solid steel tube.

First, always handle it with very clean and dry hands to avoid any possibility of etching the surface with body acids in your hands or solvent you might have on your hands. Second, this instrument can be easily damaged by a fall or setting it down too roughly. A typical cylindrical square only one foot long and four inches in diameter will weigh approximately fifty pounds! So make certain you have a good grip, and if you don't feel comfortable lifting the weight, ask for help.

THE COMBINATION SQUARE

Next to the tri-square, the combination square is the most common of precision squares used in industry. Although generally considered not quite as accurate as a fixed blade tri-square, it is sufficiently accurate for most layout work, and its versatility more than compensates for any other shortcomings. The combination square takes its name from the fact that it can have several "heads," each of which is adjustable along the length of the blade, which allows it to do a combination of tasks. Fig. 3-6 is a typical combination square set.

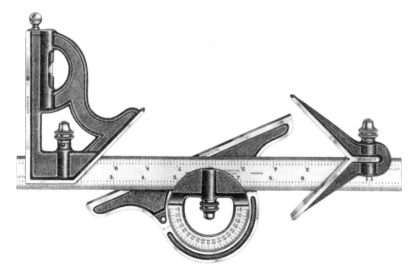

Fig. 3-6. The combination square is available with one or more separate heads which allow this instrument to perform a variety of functions. It is shown here with (from left to right) a standard head, a protractor head, and a center finder head. (Photo courtesy of The L.S. Starrett Co.)

The blade of this square consists of the equivalent of a steel rule with a groove cut in the center of one side along its length. The groove serves as a key slot to retain the adjustable head of the square, and allows the head to be locked in any desired position along the length of the blade. The blade in turn is etched with a measurement system.

The Standard Head

The standard head of the combination square acts like the wide blade of the tri-square. The advantage of this instrument is that it can be adjusted along the length of the narrow blade, thus allowing measurements in tight places. Additionally, the head can be set to correspond to any distance etched on the narrow blade. Then layout work can scribe lines directly across the end of the narrow blade and be assured they are consistently the same distance from an edge. This is especially useful for laying out hole centers which all fall in a line from a given surface.

The back side of the standard head is also an angular measuring device, but is set at a 45° angle rather than at 90°. This makes the combination square very useful for laying out cut dimensions for mitered joints.

The Protractor Head

The combination square protractor head is an angular protractor, fully adjustable both in angularity and position along the length of the narrow blade. With a full 180° of swing, this head is used for laying out lines which are not square to the reference surface. In actuality, this head can be used for layout using polar coordinates, but is more typically used for layout of cut dimensions where the miter will be some measurement other than 90°.

The reason for the constant reference to layout work is that this instrument is usually not accurate enough for inspection work. This of course depends on the circumstances of the parts being fabricated. If the fabrication is a welded construction, the combination square is more than sufficiently accurate; but if the fabrication is a precision machined part, then angular measurements should be taken with a precision protractor. We will discuss the precision protractor and angular Vernier scale later in this book.

The Center Finder

This is perhaps the combination square's crowning ability—to be able to determine a line that is square to any tangent point on a cylinder. The center finder head consists of a V-block which is set with the narrow blade of the square passing through it so that one edge of the blade passes directly through the vertex of the V. Again, this head is fully adjustable along the length of the blade.

The operator can determine the center of a cylinder by scribing any two noncollinear lines on the end of the cylinder using the center finder head. Although it is typically thought that these two lines should be perpendicular to each other, this is not absolutely necessary. Any two lines which are perpendicular to a tangent point on the cylinder will intersect at the center of the cylinder. This operation is especially useful for drilling in the end of bar stock, or center drilling for placing the stock on a fixed center in a lathe.

Gauge Blocks

Gauge blocks are the first of the instruments we will cover that are used for calibration as well as measurement. Like the surface plate, with which they are almost always used, gauge blocks are comparatively simple, but highly accurate and delicate, instruments. They consist of a block of very hard material, either hardened tool steel or chrome carbide, ground and lapped to extremely close tolerances. Also, like the surface plate, they are available in a variety of grades. Laboratory grade gauge blocks will have a tolerance as close as ±0.000001 inch (one-millionth of an inch, 1μ); toolroom grades run approximately +4μ, -2μ.

Not only are the linear measurements held to these close tolerances, but the surface finish of the ends is likewise held to the same tolerance, as is the parallelism of the opposing ends. Gauge blocks typically come in sets, ranging from only a small fraction of an inch to several inches in length, and calibrated in very small increments. Fig. 3-7 is a complete set of incrementally larger gauge blocks. By stacking the various lengths of blocks in the set, virtually any length is obtainable. Aside from making an extremely accurate measuring device, the exceptionally smooth

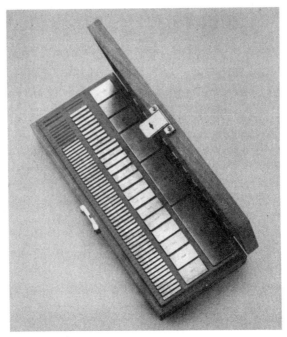

Fig. 3-7. A complete set of rectangular gauge blocks. These precision linear measurement instruments are always kept in a cloth-lined wooden box to protect them from damage when not in use. (Photo courtesy of Federal Products Co.)

surface finish allows the blocks to be "wrung" together. Wringing gauge blocks together entails holding two blocks together and twisting them a few degrees into alignment. The mating surfaces are so smooth that the blocks will literally adhere together. This provides absolute accuracy in dimensions using stacks of gauge blocks. Fig. 3-8 illustrates a typical gauge block stack.

Gauge blocks are used in essentially two capacities: measurement in conjunction with another instrument such as a sine bar or sine plate, or as a standard for checking and calibrating other instruments. We will cover the procedure for using gauge blocks in conjunction with other instruments under the sections on those instruments. For now, we will consider them as calibration instruments.

It is typical that checking an item in industry should require

Fig. 3-8. Gauge blocks can be "wrung" together to obtain virtually any possible linear measurement. Like the cylindrical square, gauge blocks are commonly used as a master standard for checking other measurement instruments. (Photo courtesy of The DoALL Co., Des Plaines, IL.)

a checking device or standard at least ten times as accurate as the item being checked. In the case of the Vernier micrometer calibrated in ten-thousandths of an inch, the checking standard needs to be at least accurate to one one-hundred-thousandth of an inch. Obviously, gauge blocks, even in toolroom grade, are even more accurate than this. Checking the Vernier micrometer for consistent accuracy is a two-step operation. First, the spindle is closed against the anvil and the scale is zeroed. This assures that the instrument is measuring from the correct baseline. Next, a stack of gauge blocks is wrung together resulting in an accurate dimensional stack that can be measured in ten-thousandths of the inch. The stack is then measured with the micrometer. If the measurement on the micrometer is equal to the dimension of the gauge block stack, then the micrometer is measuring accurately. If not, obviously the instrument has a problem—perhaps a worn lead screw—and should be repaired or replaced.

A similar procedure is used for checking the accuracy of the height gauge. The gauge blocks are stacked and placed on a surface plate, and then the height gauge probe is brought to rest on the stack and the measurement read directly from the height gauge. If the numbers don't match, the height gauge needs repair. This also explains why keeping the surface plate and the gauge blocks in clean condition and free from damage is so important; the accuracy of many other measuring instruments relies on these other two instruments.

Height Gauge

Perhaps the most common of the instruments used in conjunction with the surface plate is the height gauge. This is largely because the height gauge is one of the few instruments that cannot be used without a surface plate. Fig. 3-9 illustrates the height gauge. This instrument is, in essence, an upright Vernier

Fig. 3-9. The height gauge is actually a type of vertical Vernier caliper that uses the surface plate as one of its jaws. (Photo courtesy of Mitutoyo/MTI Corp.)

caliper with only one jaw. The surface plate operates as the remaining jaw. This allows measurements to be taken which are perfectly square to a reference plane, in this case the top of the surface plate.

The height gauge is useful as both an inspection instrument and a layout instrument. The hardened steel probe attached to the head of the height gauge is sharpened to a blade. Layout work on the surface plate can use this probe to scribe lines on the workpiece at specific distances from the resting place on the surface plate. Typically, the workpiece is coated with a blue dye, appropriately called "bluing," and lines are scribed in the dye. For example, if the center lines of a hole to be drilled are to be laid out from the edges of a rectangular plate, the plate is set on a surface plate after bluing and the height gauge is used to scribe a line representing one of the center lines. The part is then rotated to the next side and the operation repeated. Given the accuracy of the height gauge afforded by its Vernier scale, these dimensions will be accurately located within one-thousandth of an inch.

Inspection operations are accomplished in much the same manner. Additionally, the blade probe can be replaced with other styles of probes. These probes are then calibrated with the aid of gauge block stacks so that measurements taken with the height gauge are accurate. One example of this operation would be attaching a dial indicator with a probe to the head of the height gauge. The indicator would be zeroed with the aid of gauge blocks and then the part would be checked. This is especially useful for operations like checking the runout of a flat surface.

Chapter 4

Gauges

MEASUREMENTS, ESPECIALLY REPETITIVE measurements, do not usually require adjustable instruments such as the caliper or micrometer. This is especially true for repetitive measurements in checking operations where the part being measured is either acceptable, or must be discarded or repaired. This type of measurement is almost invariably associated with checking operations, and employs a measurement "standard" called a "gauge."

Gauges are any of the various instruments which are made to check a given measurement and ascertain if the part being checked is made within the specified tolerances. Therefore, the gauge is made to very exacting tolerances and used as a standard against which the part is compared. If the part does not fit the gauge, then it must be repaired (if this is possible), or it must be discarded. Gauges are therefore the essential element in fabrication of parts for use in interchangeable manufacturing.

Gauges may be either standard instruments or specially made instruments to fit a particular measurement application. For example, standard gauges are manufactured for measuring items like threads. This stands to reason since threads are all manufactured to certain preestablished sizes and shapes. The number of threads per inch, the height of the points, the depth of the valleys, and the pitch diameter are all standard items regarding threads that should never vary. Therefore, a thread gauge made for a 1/2-13 external thread should check any such thread made in any manufacturing facility.

This same principle applies to all standard items that can be checked with a gauge. However, specially made gauges are equally frequent in industry for making repetitive measurements of parts. Any gauge that can be purchased for checking a standard measurement can also be made or modified to check a nonstandard measurement. Additionally, gauges can be made for checking special situations on products specific to one manufacturer. Perhaps the most common application of special gauges is checking the size and shape (or contour) of complex shaped parts. For example, suppose a manufacturer makes the blades for the screw (propeller) of a large ship. These blades must be exactly the correct size, shape, and contour so they will not vibrate during use. Vibration would not only be a nuisance, by making the ship rattle, but would shorten the life of the screw by causing stress cracking, and would damage the shaft and bearings that operate it. However, the screw blades are curved, twisted, and warped in all axes. To check the contour of these blades, the manufacturer could make a special gauge with templates cut to the exact contour of the blade, and place the blade in the gauge. Assuming the gauge is made to the exact tolerances of the design, the blade will fit in the gauge if the blade is correctly made. If the blade is not correctly made, it would have to be corrected, or scrapped and recast.

Whether the gauge is standard or specially made, this is essentially the principle of the gauge: the part either fits or it doesn't. Gauges are typically not adjustable, as they are intended for one specific operation and nothing else. This may seem like it would entail many separate tools to check all of the measurements in any given assembly—and honestly it does. The advantage to gauges is the speed at which they can be applied. This is absolutely essential in repetitive operations typical in mass manufacturing.

Checking a screw thread with a thread micrometer is possible, but a thread gauge takes only a small fraction of the time to apply, and with reasonable accuracy. If the operator must check a screw thread on a manufacturing line every 37 seconds, the micrometer would be nearly impossible to use. Another situation would be the above-mentioned example of the ship screw blade. This item would be virtually impossible to check with most standard measuring instruments, except for something like a Coordinate Measuring Machine or laser interferometer. These latter instruments would work very nicely, and are frequently applied to this type of measuring, but they are also very costly to purchase and maintain, whereas a gauge would work equally well at a fraction of the cost.

In this chapter, we will only cover standard purchased gauges, as it would be virtually impossible to include all of the possible specialty gauges used in industry. Keep in mind, however, that all of the principles used in standard gauges can be applied to making specialty gauges.

Plug and Ring Gauges

Since standard gauges are used for checking the accuracy of common industrial operations, it will come as no surprise that most of the features checked are fairly simple ones. Perhaps the most common application of standard gauges is for checking the diameter of holes and shafts. As a rule, most holes, regardless of internal features, are made with drills, punches, reamers, or counterbores of a set and standard size. Likewise, most of these holes will receive shafts, fasteners, dowels, or other diametrical items that must fit inside of the hole with consistent clearance or interference.

The gauges for these features are essentially a highly accurate representation of the mating component. A hole will be checked with a shaft shaped gauge, called a "plug gauge," and a shaft will be checked with a highly accurate internal diameter, called a "ring gauge." However, placing a highly accurate shaft in a hole will only tell the operator that the hole is not too small. Any diameter over the size of the gauge will still accept the gauge, but not tell the operator anything else about it. To compensate for this shortcoming, plug and ring gauges typically are made with two checking surfaces, either shafts or holes depending on the type of gauge.

These double-ended plug gauges or ring gauges are called "go/no-go gauges." The object of these gauges is to provide for two

Fig. 4-1. Cylindrical go/no-go plug gauge used for checking the diameter of holes. Note the notches on the check are only on one end; this is to facilitate easy identification of one gauge end from the other. (Photo courtesy of Carr-Lane Mfg. Co.)

check surfaces, both of which are just a tiny fraction beyond the upper and lower tolerances of the hole or shaft. Fig. 4-1 is a typical go/no-go plug gauge.

Straight Plug and Ring Gauges

Let's consider how these gauges work for a moment. Suppose a hole to be checked is specified on the design prints as 0.500 ±0.010. The smallest the hole can be and still meet design requirements is 0.500 - 0.010 = 0.490, and the largest the hole can be and still meet design requirements is 0.500 + 0.010 = 0.510. Therefore, the plug gauge to check this hole should have one shaft that is just slightly smaller than 0.490, or about 0.4899, and the other end just slightly larger than 0.510, or about 0.5101. The small shaft is the "go" end, and is inserted in the hole first. If this end is small enough, it should fit in any hole falling within the tolerance limits. If it doesn't, the hole is too small and no further checking is required. Next, the large end of the plug gauge is inserted, or attempted to be inserted, in the hole. This end is the "no-go" end, as it is larger than the hole should ever get; and if the hole is made correctly, it will not fit into it. If the no-go end does fit into the hole, obviously the hole is too big and the part is probably scrap. If the go end of the gauge will not enter the hole, it is usually possible to open the hole up with additional machining, and therefore redeem the part in process; but filling in a machined hole is usually more costly than scrapping the part due to labor costs.

The application of plug and ring gauges is very common, especially in relation to straight holes and shafts. Recall that in Chapter 1 we discussed fits and clearances. This is the condition where a shaft that will be assembled into a hole must either clear or interfere by an exact amount. These conditions constitute slip-fits and press-fits in industry. They are not arbitrary amounts, but very specific amounts of clearance or interference to obtain a correct motion in clearance conditions or a sufficient hold in press-fit conditions.

For example, many machine parts are held in location by dowels. Screws hold the components of the machine together, but dowels hold the pieces in location relative to each other. The dowels will be press-fit into one of the components and slip-fit into the mating component. To assure that these components will assemble properly, with the correct amount of interference on the press-fit side and the correct amount of clearance on the slip-fit side, the mating holes are checked with a straight plug gauge.

On the other hand of the operation, the dowel manufacturers will take periodic samples of their dowels and check them for correct diameter with a ring gauge. Dowel pins are hardened and then ground to an exact diameter on a machine called a centerless grinder. Periodic checking of the dowels assures the dowel manufacturers that they are shipping products that will not be rejected by the customer because of errors in the grinding process.

Taper Plug and Ring Gauges

The procedures described so far have been based on the assumption that the hole or shaft under consideration is a straight diameter with no contour. There are, however, instances when a hole must either have internal contour, or a shaft external contour, such as threads or tapers. First, let's consider the taper of holes or shafts.

Obviously, the type of go/no-go gauge used for a straight hole or shaft could not be used to check a taper. The taper plug gauge is therefore a single-ended instrument rather than a double-ended instrument. The shaft of the taper plug gauge is ground and polished to a very accurate representation of a tapered shaft, with the diameter, taper, and concentricity of the large and small diameters of the taper all made exactly. However, how can the operator tell if the instrument is reading the correct values?

To accommodate the diametrical measurement, the gauge is etched with two lines (or provided with two ground steps) at the largest and smallest diameters of the large end of the taper. When the instrument is inserted into the tapered hole, the large diameter line should remain outside of the hole and the small diameter line should be within the hole. Next, to check for correct taper, the operator must gently try to wiggle the instrument in the hole. If the instrument can move from side to side in any direction, the taper is incorrect. Movement at the large end of the taper means the taper is too steep. Movement at the small end of the taper means the taper is too shallow.

This procedure will not tell the operator if the taper is misshaped between the large and small ends of the taper. To accomplish this, the gauge is coated with a light coat of bluing and given a slight twist on the mating surface. When the instrument is extracted, it is checked for marks in the bluing. If the rub marks are even, and evenly distributed, the taper is properly shaped. Conversely, if the bluing has rubbed off unevenly or sporadically, the taper is misshaped.

The taper ring gauge works in very much the same, albeit opposite, manner as the taper plug gauge. The exception to this rule is that the bluing application for checking the shape of the taper will go on the part instead of the gauge. In other words, bluing for checking the shape of a taper with a gauge will always go on the shaft, regardless of whether this is the gauge or the part.

THREAD PLUG AND RING GAUGES

Thread plug gauges are double-ended go/no-go instruments similar to straight hole plug gauges. The gauge ends are of different lengths for easy identification, with the long end being the go end and the short end the no-go end. The threaded ends of the instrument have a chip-groove cut along them to eliminate any chips that may be present, but this is only a precaution and should not be used for cleaning excess chips from the threads. Fig. 4-2 is a thread plug go/no-go gauge.

The go end of the instrument is engaged in the threaded hole and screwed all the way to the bottom of the thread to check it. Excessive binding of the go end is indicative that the thread is too small and needs to be retapped. The no-go end is then engaged in the thread as far as it will go, which, if the thread is formed

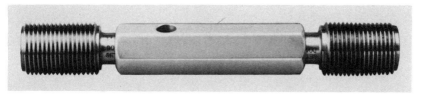

Fig. 4-2. The thread go/no-go gauge. Care must be taken never to force this instrument into a threaded hole, as it could damage the instrument, and will invariably give an inaccurate reading. (Photo courtesy of The DoALL Co., Des Plaines, IL.)

correctly, should be no further than flush with the surface of the part. If the no-go end engages below the surface of the part, then the threaded hole is too large and the part needs to be repaired or scrapped. Repairing a threaded hole will usually entail opening it up to a larger diameter and installing a thread insert of the originally intended diameter.

The thread ring gauge strongly resembles the straight ring gauge, with a few possible exceptions. The thread ring gauge go and no-go gauges are separate rings, typically mounted in a single holding bracket, with the openings across from each other. The part is first threaded completely into or through the go side of the gauge. Next, the part is engaged into the no-go side of the gauge, but should not travel further than one and one-half turns into the gauge maximum. Generally, the surface of the gauge is stamped with identification to differentiate the go side from the no-go side. On some gauges of this type, the no-go side of the gauge will have a groove cut around the outside of the gauge. Some thread ring gauges are manufactured with each gauge comprised of three segments connected with set screws to allow for small adjustment in the internal diameter of the gauge.

Snap Gauges

The snap gauge is a close relative of the caliper, but is designed specifically for repetitive checking measurements, typically of external diameters. There are many varieties in the configuration of the snap gauge, but all are similar in function. They will have a C-shaped frame like that of a micrometer, with a fixed anvil on one side of the frame and two adjustable anvils

Fig. 4-3. The snap gauge is a convenient tool for checking external dimensions such as diameters. Because the adjustable anvils are independent of one another, the snap gauge is actually an adjustable go/no-go gauge. (Photo courtesy of The DoALL Co., Des Plaines, IL.)

on the remaining side of the frame. The two adjustable anvils are set to the upper and lower limits of the part being checked, the largest dimension being set at the anvil nearest the exit of the frame. Fig. 4-3 illustrates a typical snap gauge.

In this fashion, the snap gauge is another of the family of go/no-go gauges. If the diameter of the part being checked is sufficiently small, it will pass freely through the first set of anvils as the go check. If the part diameter is sufficiently large, it will not pass beyond the second set of anvils acting as the no-go check.

The adjustable anvils of the snap gauge are first set to the correct gap using a gauge block stack of the required height. Each anvil is then locked in place and rechecked against the gauge blocks. As with all go/no-go gauges, the surfaces of the gauge and the surfaces to be checked must be kept clean and dry at all times to ensure accurate checking.

Feeler Gauges

Feeler gauges are fairly simple measurement instruments for taking comparatively rough measurement checks of gaps or small simple contours. These instruments are in essence a template of a standard feature. In the broadest terms, feeler gauges are also go/no-go gauges in the sense that the template will either fit the contour satisfactorily or it will not fit. Due to the rough measurement characteristics and the comparatively simple nature of these instruments, we will only touch on each of them briefly, but keep in mind that each of these tools can be an invaluable time-saving aid for checking work in progress in the machine shop.

THICKNESS FEELER GAUGES

The thickness feeler gauge is a set of small instruments consisting of either flat blades or rods made to a high tolerance, and in progressively larger thicknesses. They are used exclusively as a gap check instrument. The gauge is inserted into a gap to determine if the gap is the correct size. Again, this is essentially a go/no-go situation, because the gauge will either fit or it will not. By using progressively larger or smaller gauges in the set, the size of the gap can at least be determined within a certain range.

Generally used in automotive operations to adjust the gap within the firing end of a spark plug, the thickness feeler gauge finds many applications in industry. One application in particular is checking contours of parts by placing a rod-type feeler gauge between the part and a specially made template representing the part's contour, which is offset from the part forming a predetermined gap. In this application, it should be obvious that standard gauges can be successively used in conjunction with specialty gauges.

RADIUS GAUGES

Radius gauges somewhat resemble a set of flat blade thickness feeler gauges, with either concave or convex radii formed in the end of the blade. These gauges are templates of incrementally larger radii for taking checks on such items as external rounded corners or filleted internal corners. Fig. 4-4 illustrates a set of radius gauges and two different sets of thread gauges.

THREAD GAUGES

Like the radius gauges, thread gauges come in sets held together on a common holder. These gauges are used for operations such as checking the development of work in progress of a threaded shaft being cut on a lathe. The thread gauge is used to check the pitch diameter of a particular thread form. They can be used on external threads, or on internal threads where the diameter of the hole permits.

ANGLE GAUGES

Angle gauges should not be confused with angle gauge blocks, which we will discuss in detail in the next chapter. The angle gauges mentioned here are again similar to the flat blade thickness feeler gauge, but the end of the blade is cut to a specific angle in relation to the side of the blade. These instruments also come in collected sets like those of the radius gauge shown in Fig. 4-4. The increments of the angle gauges are fairly large, with the smallest graduation being one degree. While this is sufficient for in-progress operations requiring only sight check accuracy, they are not substitutes for angle gauge blocks and a surface plate.

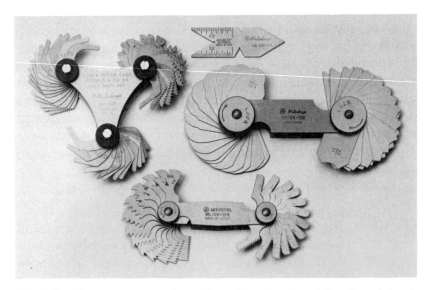

Fig. 4-4. Shape gauges such as the radius gauge and the thread gauge provide a quick means of checking standard items with reasonable accuracy. Note the radius gauge has internal radii on one end and external radii on the other. Thread gauges will cover the entire range of thread sizes including standard, fine, and extra fine threads. (Photo courtesy of Mitutoyo/MTI Corp.)

BALL AND DIAMETER GAUGES

The last of our template style gauges is the ball and diameter gauge. This instrument, sometimes called a wire gauge, is essentially a steel plate pierced with a series of incrementally larger holes. Each hole is a specific diameter and is held to a relatively close tolerance. This gauge is used to check the diameter of small spheres or rods, such as a length of wire, for correct diameter. Another of the uses of this instrument would be to ascertain the diameter of a standard drill bit, in the event that the size normally etched on the shaft is marred due to slippage in the drill chuck.

Cutter Clearance Gauge

The cutter clearance gauge is a fully adjustable gauge used for checking the backdraft angle of milling cutters. It consists of three feet—one fixed and one adjustable toward or away from the fixed foot, and a third foot that is adjustable up/down and angularly. The first two feet are adjusted to rest on the crest of two teeth, which would be the first and third in any group of three. The remaining foot is adjusted to the desired angle relative to the plane of the other two feet and brought to rest on the back side of the cutter tooth. See Fig. 4-5.

This instrument is used to determine if the cutter is ground correctly, or if the particular backdraft angle of a given cutter will perform correctly with the material to be machined with the necessary speed and feed rates. The backdraft angle of a milling cutter will greatly determine the speed and feed rate at which any given cutter will perform without chattering and thereby marring the work surface.

Drill Point Gauge

The drill point gauge is a small steel rule with an angular template attached to the side for checking the draft angle on the tip of a drill bit. Sharpening the bit is frequently a manual hand-held operation, often relying on the expertise and sight of the operator. The drill point gauge is used by the operator for checking the draft angle periodically during grinding operations. The importance of this operation cannot be stressed enough. The

Fig. 4-5. The cutter clearance gauge is used to determine the correct amount of depth and backdraft of cutter blades on a milling machine cutter. This is very important for proper operation of the machine and correct surface finish on the workpiece. (Photo courtesy of The L.S. Starrett Co.)

symmetry of the drill bit point is absolutely imperative to reduce the tendency of the drill bit to wander across the workpiece prior to embedding. If the point of the drill bit is not exactly centered, then the wander of the bit will be increased. Fig. 4-6 illustrates the drill point gauge.

Fig. 4-6. The drill point gauge is used to check the draft angle of the cutting point of a drill bit. Insufficient or excessive draft angle will cause the bit to cut improper, uneven, or out-of-tolerance holes, and will also cause excessive wear to the tool. (Photo courtesy of The L.S. Starrett Co.)

Countersink Gauge

Countersinks typically must be made to a specific size to allow the flat head fasteners they are to receive to be flush with the surface of the part. However, checking this contour and depth is awkward. To accommodate this situation, the machinist can use a special countersink gauge.

The countersink gauge consists of a spindle with an appropriately angled cone on the end, an outer housing, and a specially calibrated dial indicator. The conical end of the spindle is manufactured to one of several standard countersink angles, and the tip is hardened to reduce wear. As the instrument is forced down into the countersink, the cone of the spindle will self-center in the countersink. The spindle is pushed up through the housing and activates the dial indicator on the top of the housing. The housing itself acts as a stop when it comes to rest on the surface of the workpiece. When the housing comes to rest, the reading on the dial indicator will indicate the diameter of the countersink.

Universal Precision Gauge

The universal precision gauge has nearly unlimited applications, but it is most frequently used to set up the cutter height for the first pass of a planer or shaper. The instrument itself consists of two forged steel, triangular blocks, one larger than the other, which connect by means of a clinch nut and T- or L-shaped slot in the larger block. The smaller of the two blocks can be adjusted up and down along the mating incline, thus increasing or decreasing the distance between the base surface and the top of the gauge. Fig. 4-7 illustrates a universal precision gauge.

The height of the gauge can be set with the aid of a caliper, micrometer, height gauge, gauge blocks, or other measuring device, and then clamped in position. Aside from serving as a height reference, the universal precision gauge can have any one of several other measuring instruments attached to the top of it, such as dial indicators, scribers, or, most commonly, an additional height standard consisting of a length of ground steel shaft held to a specific length tolerance.

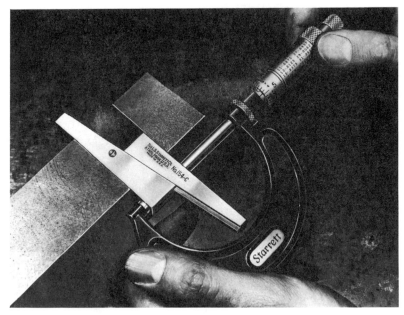

Fig. 4-7. The universal precision gauge operates on two opposing in-clined planes with a locking mechanism. The addition of a cylindrical extension (not shown in this illustration) allows the instrument to ob-tain a much larger range. (Photo courtesy of The L.S. Starrett Co.)

Adjustable Gauges

Some gauges, like the universal precision gauge and the snap gauge, are adjustable over a limited range to fit a variety of applications. Two other common adjustable gauges are the "small-hole gauge" (or "split-ball gauge") and the "telescoping gauge." Each of these types of gauges is typically found in sets in order to cover a reasonable range of distances. Fig. 4-8, for exam-ple, is a set of five split-ball gauges covering a range of 0.125″– 0.500″.

They operate by point contact on an internal surface such as a hole, by being adjusted outward with the knurled handle until they contact. After adjusting the instrument, the operator re-moves it and checks the outer point distance with a micrometer or caliper. Aside from checking small hole diameters, these in-struments are useful for checking other small internal features such as the channel shown in Fig. 4-9.

Fig. 4-8. Split-ball, or small hole, gauges typically come in sets to cover a wider range of measuring. The distance across the points of the split-ball is adjusted by turning the knurled handle to drive a wedge between the halves of the gauge. (Photo courtesy of The L.S. Starrett Co.)

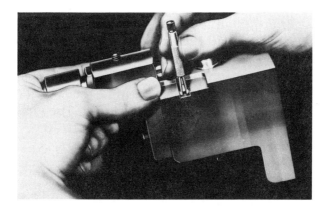

Fig. 4-9. Aside from checking small hole diameters, these instruments are useful for checking other small internal features such as the channel shown here. (Photo courtesy of The L.S. Starrett Co.)

Fig. 4-10. Application of a telescoping gauge for checking the internal diameter of a workpiece during lathe operations. (Photo courtesy of The L.S. Starrett Co.)

Similar in operation to the split-ball gauge, the telescoping gauge shown in Fig. 4-10 is used to check internal diameters of greater diameter. Again, the instrument is adjusted out to point contact in much the same way a standard non-Vernier internal caliper is used, and once removed from the workpiece, it is checked with a Vernier caliper or micrometer. The typical set of six telescoping gauges will cover a range of 5/16″–6″.

Angular Measuring Devices

AFTER LINEAR AND SQUARENESS MEASUREMENTS, angularity is the most common measurement concern in industry. Angular measurements are involved in everything from setting a mitered joint to calculating the framework used in bridge construction. In this chapter, we will cover the instruments used in measuring angles (except for right angles, which were covered in the section on surface plate instruments). It is important that the reader have at least a rudimentary understanding of trigonometry, as the use of some angular measuring instruments will require certain basic calculations.

For the most part, angular measurements in industry are made using degrees, regardless of whether other measurements are in English units or Metric units. Radians and grads are typically used more in design and scientific calculation than in industrial manufacturing. Consequently, all of the considerations in this chapter will be in degrees, and fractional portions will either be in minutes and seconds, or in decimal degrees.

Angle Gauge Blocks

Angle gauge blocks are similar to square or rectangular linear gauge blocks, except that they are machined in progressively

Fig. 5-1. A complete set of angle gauge blocks will accurately measure any angle because of the ability to stack them in either direction. This particular set is graduated in 1/4" increments, although other sets are available in 1" increments. (Photo courtesy of Mitutoyo/MTI Corp.)

larger angles instead of longer lengths. Their end surfaces are ground, polished, and lapped, so they too can be wrung together to build up stacks representing different angles. However, since the top and bottom surfaces are at angles to each other, the direction they are wrung together is critical to obtain the correct angle. Fig. 5-1 illustrates a complete set of angle gauge blocks, incremented in one-quarter-degree steps. Some complete sets are capable of making angular measurements in increments of one second accuracy! The blocks themselves may be lapped to an accuracy as close as one-quarter second ($1° \div 60' \div 60'' \div 4 = 0.0000694°$).

Such a set of angle gauge blocks incremented in seconds will therefore measure any angle between 0° and 99° in one second increments; or, in other words, any one of 356,400 different angles up to 99°, which would constitute all of the blocks included. This number of angles represents 99 degrees multiplied by 60 minutes for each degree, and multiplied by 60 seconds for each minute. But just how many blocks would there have to be

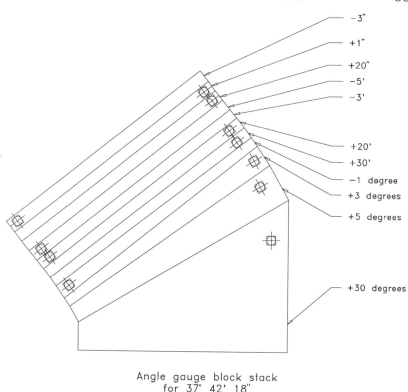

-3"
+1"
+20"
-5'
-3'
+20'
+30'
-1 degree
+3 degrees
+5 degrees
+30 degrees

Angle gauge block stack
for 37° 42' 18"

Fig. 5-2. Illustration of the application of angle gauge blocks. Note the small symbol on the end of the block to indicate which is the large end.

in the set to measure such a vast number of angles? Surprisingly, the answer is 16 blocks. The reason is that the angle of each block, or set of blocks, can be either added to or *subtracted from* the angular value of any other block or set of blocks. Therefore, the only blocks necessary are six blocks in even degrees—1°, 3°, 5°, 15°, 30°, and 45°; five blocks in even minutes—1', 3', 5', 20', 30'; and five blocks in even seconds—1", 3", 5", 20", 30".

Let's examine an example to see how this is accomplished. Fig. 5-2 shows an angle gauge block stack representing a specific angle. Building a stack of angle gauge blocks is a bit like doing long division: the largest numbers are divided first and then the remainder is divided. For the sake of convenience, it is best to wring together the largest parts of the angle first and work down from there. Degree increment blocks are put together first, followed by minute blocks, and finally second blocks. In the illustration, we see the desired angle is 37° 42′ 18." The largest block first is therefore the 30° block. Next, the 5° block would be wrung onto the stack for 35°. Now we are faced with two degrees left to add, but only a 1° and a 3° block. The solution here is to add the 3° block, and then add the 1° in *reverse* position. In other words, we are adding three degrees and subtracting one degree. The remaining values for minutes and seconds are progressively accomplished in the same manner.

Like parallel gauge blocks, angle gauge blocks are for the most part a surface plate instrument. They can also be used on other reference planes, such as a milling machine table, but in either case they should be handled with extreme care to avoid marring the lapped surfaces of the angles. For surface plate applications, the blocks can be placed on a workpiece with an angular surface, and the blocks read with a height gauge or dial indicator to check for accuracy of the angle. If the workpiece is quite small, it can be placed on top of the blocks, which in turn are resting on the surface plate, and the check can be made directly on the workpiece.

The Protractor

Angle gauge blocks comprise a highly accurate instrument, but they are generally restricted to use on a surface plate, and they are primarily an angular *surface* instrument. Layout work typically involves scribing *lines*; and scribing lines is generally faster and easier with a "protractor." This device was touched on briefly during our discussion of the precision square when dealing with the protractor head of the combination square. Both the protractor head and the fixed protractor work in the same manner and are fairly simple instruments to read and understand. We will cover the fixed protractor, but keep in mind that adjusting the protractor head of the combination square is accomplished with the same principles.

In Chapter 1, we covered the definition of the degree as it relates to part of a circle or an arc. The protractor is simply a representation of an arc, with graduations etched at each degree of angularity. Fig. 5-3 is an illustration of a typical protractor.

Fig. 5-3. The precision universal protractor employs a standard protractor with one degree graduations. This instrument operates on the same principle as a student's or draftsman's protractor. (Photo courtesy of The L.S. Starrett Co.)

Note that the graduations at zero are directly across from each other. This forms a baseline from which all other angles are measured. The baseline is aligned with one of the legs of the angle, and the vertex of the angle is exactly aligned with the tick mark at the middle of the baseline. The remaining leg will then intersect the arc of the protractor, and the measurement can be directly read from the graduations.

The industrial protractor used in most machine shops is a slightly modified version of this same instrument, with an extension leg to facilitate scribing lines further from the arc of the instrument and with greater accuracy. The leg of the industrial protractor is connected to the protractor arc by a pivot along the baseline. A small tang with an etched tick mark travels along the arc and indicates the graduations of any given angle. The leg then extends below the baseline allowing easy access for scribing along its edge.

THE ANGULAR VERNIER

While the industrial protractor is fairly easy to read and use, with precision to one degree, any application requiring accuracy greater than one degree will require a special instrument called a Vernier bevel protractor. This instrument employs a regular protractor that encompasses a full circle, and has the added feature of an angular Vernier scale. Like the linear Vernier used on calipers and micrometers, the angular Vernier scale uses two opposing sets of graduations which have slightly different spacing. With the addition of the angular Vernier scale, the protractor is now capable of measurements to 1/12 of a degree or, in other words, within five minutes.

Fig. 5-4 illustrates the angular Vernier scale. The main scale etched on the circumference of the protractor is graduated in whole degrees just like any other protractor. The secondary scale is a short segment of arc just below the main scale with 12 graduations which cover the same span of arc as 23 graduations on the main scale. Each of the Vernier graduations is therefore 1/12 smaller than those on the main scale. Reading the angular

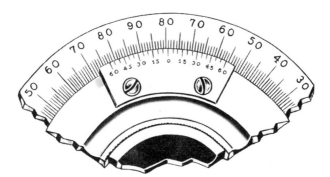

Fig. 5-4. Closeup of the angular Vernier scale, with a reading of 50°20′. This device is used on industrial protractors as well as the drafting machines used by industrial designers and engineers. (Photo courtesy of The L.S. Starrett Co.)

Vernier is accomplished in a manner similar to the linear Vernier, except that the scale can be read from either direction depending on the direction the angle is being measured from.

First, the angular degrees are read at the point of coincidence of the zero line on the Vernier and the degree graduation on the main scale. If these points coincide perfectly, then the angle is a number of even degrees. If, however, they do not perfectly coincide, then the next step is to look for the Vernier graduation which exactly coincides with *any* tick mark on the main scale. Once this is found, the value of the Vernier scale is added to the main scale to determine the angle. However, since the Vernier protractor scale and the main scale can both be read on either the left or right side of the zero, then the Vernier graduation that coincides with a main scale graduation must be one on the side of the Vernier zero line in the same direction the angle is being measured from. In other words, if the angle is being measured with main scale graduations which are progressing counterclockwise, then the Vernier is read on the left side of the zero line. If we refer to our illustration, we see that since each Vernier graduation equals 5′, we have an angle of 50° 20′.

Some angular Vernier scales are even more accurate than the one described here. For example, the angular Vernier used on a drafting machine in the design room can have an accuracy of one minute rather than 5′. However, the reading process is the same, and it is best to refer to the owner's manual to be certain what degree of accuracy the instruments are designed for.

The Sine Bar

At this point, it will be necessary to understand at least rudimentary trigonometry, because the remainder of the instruments in this chapter all rely on some form of trigonometric calculations in order to use them. The sine (pronounced like the word "sign") bar is a straight, flat bar of polished steel with a notch cut in each end. Attached in the notches at each end are two polished steel bars of exactly the same diameter, which are exactly parallel to the top and ends of the sine bar. The sine bar rests on these steel rods, which in turn rest on a reference plane, but allow the plane on one end of the sine bar to be at a different

Fig. 5-5. The 5- and 10-inch precision sine bar used for measuring and setting up work at accurate angles. This instrument is always used in conjunction with a surface plate and gauge block set. (Photo courtesy of Mitutoyo/MTI Corp.)

elevation than the elevation of the reference plane on the other end of the bar. Fig. 5-5 illustrates two sine bars of different sizes.

The sine bar is a highly accurate angular measuring device, which almost invariably must be used in conjunction with a surface plate and a set of gauge blocks. The surface plate forms the reference plane for one end of the sine bar, while the gauge blocks resting on the same surface plate form the reference plane for the other end of the sine bar. When the sine bar is resting completely on the surface plate, the top of the bar is exactly parallel to the surface plate. The addition of the gauge blocks raises one end of the sine bar, forming an angle to the surface plate. Since the gauge blocks are made to such an exact tolerance, the size of the angle can be easily calculated.

Calculating the degree of the angle formed by the sine bar requires knowing something about the particular sine bar in use, and the dimension of the gauge block stack used to set up the angle. Illustrated in Fig. 5-6 is a typical angle setup using a sine bar and gauge block stack. With this information, we now can calculate any angle based on any given stack of gauge blocks. Sine bars are typically manufactured as 5-inch, 10-inch, or 20-inch-long sine bars. This specification distance is the distance *between the centers of the parallel rods* on which the sine bar sits. For the sake of simplicity, we are using a 10-inch sine bar in our illustration and calculations.

The three components now form the sides and angles of a right angle triangle, which can be easily calculated. The sine bar

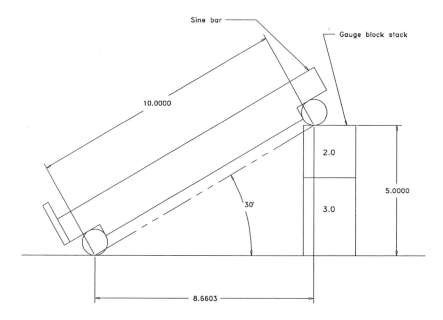

Sine bar

Gauge block stack

10.0000

2.0

5.0000

30'

3.0

8.6603

Fig. 5-6. Technique for setting up a sine bar with a gauge block stack.

forms the hypotenuse of the triangle, the surface plate forms the angle's adjacent side, and the gauge block stack forms the side opposite the angle being calculated. It may seem that we are belying the opposite side of the triangle by using the height of the gauge block stack instead of the distance from the surface plate to the center of the rod because the hypotenuse is along the centers of the rods. However, the length of the tangent points of the rods resting on the surface plate and gauge blocks will also remain at 10 inches along the hypotenuse, regardless of how the sine bar is rotated. Simple proof of this is that if we added the radius of the rod on the gauge block stack to the one end, we would have to add it to the rod on the other end to use the same points in space. The line would move up, but the length would not change.

Calculating the angle formed by the sine bar is now a simple matter of finding the sine of the angle formed between the sine bar and the surface plate. The sine of any angle is equal to the length of the side opposite the angle divided by the length of the hypotenuse. It should become immediately apparent that this is the advantage to using a 10-inch sine bar—because the sine of the angle is automatically 1/10 of the gauge block stack.

$$\text{sine } a = \frac{\text{side opposite}}{\text{hypotenuse}}$$

$$\text{sine } a = \frac{\text{gauge block stack}}{\text{length of sine bar}}$$

$$\text{sine } a = \frac{\text{gauge block stack}}{10}$$

For example, if the angle to be set up is 30°, we will need a gauge block stack ten times the sine of 30°. A quick glance at a trig table (or a quicker calculation on nearly any hand-held calculator) will reveal that the sine of 30 is 0.5000. Therefore, we will need a gauge block stack 5 inches tall to achieve a 30° angle. If the angle is unknown, the gauge block stack is used as the known. If the gauge block stack is 3.3728, then the sine of the angle is 0.33728. Again, we can look through a trig table to find this number under the sine column (or we can use the inverse function on most calculators) to determine that a sine of 0.33728 is equal to an angle of 19.7112°. From our discussion (in Chapter 2) of converting decimal degrees to degrees, minutes, and seconds, we arrive at 19° 42′ 40.3200."

Now let's consider some applications of the sine bar. First, the sine bar is predominantly a checking instrument, then a layout instrument; but it would not normally be used to set up and machine a part. A checking operation might be determining the accuracy of a taper. The part is attached or rested on the sine bar, and the angle of the taper is set up on the sine bar using the appropriate gauge block stack. Once this is done, a dial indicator on a stand or height gauge is zeroed at one end of the taper and passes along its length. If the part is made correctly, then the dial indicator should continue to read zero. If the indicator varies, then the angle of the taper is incorrect.

At this point, a word about the angles of the sine bar is in order. It is generally not good practice to use the sine bar to measure angles from the surface plate in excess of 60°; less than 45° is preferable. This is because as the angle increases, the length of the side opposite the angle, i.e., the gauge block stack, requires progressively smaller increases for each unit of angularity increase. In other words, as the angle of the sine bar passes 60°, a very small increase in the gauge block stack will result in a comparatively large change in the angle. For example, to change the angle of the sine bar from 30° to 35°, the gauge block stack must increase from 5.0000 inches to 5.7358 inches, or an increase of 0.7358. However, to change the angle of the sine bar from 60° to 65°, the gauge block stack must only increase from 8.6603 inches to 9.0631 inches. The increase in the first case is 14.7%, whereas the increase in the latter case is only 4.7%.

The result of this phenomenon is that the accuracy of the sine bar decreases dramatically after 60°. As the sine bar approaches 90°, a difference of a few thousandths will greatly change the angle. To avoid this condition, it is necessary to take advantage of the complementary angle to the angle under consideration. This entails either turning the part on the sine bar so it rests on another side, or attaching the part to an angle plate, aligning the side of the part, and turning the angle plate on its side, and then measuring the part.

The Sine Plate

Closely related to the sine bar is the sine plate. The sine plate is constructed very much like the sine bar except that it is wider and provides a means of attaching the workpiece to the sine plate. The attachment may take one of two forms: it is either a series of tapped holes to apply a clamp, or a magnetic surface where the magnet can be turned off and on. Fig. 5-7 illustrates the sine plate.

Due to the large size of the sine plate, it is possible to do light machining of the workpiece while it is attached to the sine plate, but this is not the best situation, and a sine vise or adjustable vise is a better machining instrument. In the event that machining has to be done on the surface plate, it will most likely be final operations such as light grinding. Most workpieces will be machined in a fixture for the bulk of material removal, and only then will finishing operations possibly be done on the sine plate. For more on this procedure, refer to *Basic Fixture Design* (also published by Industrial Press). The sine plate is best used on a

Fig. 5-7. The standard or simple sine plate, used for setting up work-pieces with a straight angular surface. Note the holes in the top of the plate to accommodate attachment of the workpiece. (Photo courtesy of Mitutoyo/MTI Corp.)

surface plate in the same manner as the sine bar. These two instruments are virtually interchangeable; only the size of the workpiece will determine which is the preferable tool.

The sine plate is also equipped with a flanged stop on one end to assist in aligning the workpiece with the plate. The stop will also assist in keeping the workpiece from moving on the sine plate during those times when machining on the plate is neces-sary. A variation of the sine plate is the hinged sine plate; this instrument provides a base plate under the sine plate for attach-ment to a machine table.

The Compound Sine Plate

The compound sine plate is essentially two simple sine plates stacked on top of one another at right angles to each other. The lower sine plate is hinged to a base, and the upper sine plate is hinged to the lower sine plate. Fig. 5-8 illustrates the compound sine plate. Note that the top sine plate is fitted with two rest stops. These stops are used to help align the workpiece so that it is square to the edges of the sine plate. When working with com-pound angles, this is especially important since any deviation from squareness to the plate will result in an angle that is run-ning at the right angle, but in the wrong direction relative to the sides of the workpiece.

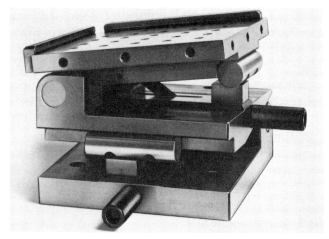

Fig. 5-8. The compound sine plate is comprised of two simple sine plates mounted on top of each other at right angles. This instrument is used to set up angles which do not necessarily run in a straight plane across the part. (Photo courtesy of Mitutoyo/MTI Corp.)

Let's consider an application of the compound sine plate. First, with the simple sine plate, the angle would run across the workpiece in a straight line from one side of the piece to the other. For example, if a block of steel were machined with an inclined surface on it running from one side to the other, the simple sine plate would be a sufficient tool. However, if the line of incline were to run from one corner of the block to the opposite corner, then the compound sine plate would be necessary.

The angle in this case is sloping across the block, and simultaneously from side to side. Such blocks are commonly made as locator blocks in the manufacture of fixtures which will hold sheet metal parts with complex surfaces such as those found in automobiles or aircraft. The block will most likely be numerically control machined (NC) for metal removal, but grinding could be done on the compound sine plate.

Each angle of the compound surface plate is set independently with a set of gauge blocks. The order of setting the angles is inconsequential, with either the lower sine plate or the upper sine plate completely set before proceeding to set the angle of the remaining sine plate. The procedures for setting the angles of the compound sine plate are the same as those for setting the angle of the sine bar.

The thing to remember about using any of the sine bar or sine plate related instruments is that the surface under consideration is to be brought parallel to the surface plate or machine table. Even though this surface is technically the angular surface, it must be made parallel to the surface plate in order to measure it, or parallel to the machine table to machine it. In the case of measuring the workpiece, the object is to set the angles to those specified on the design prints, and run a dial indicator over the surface in question. In the case of using the compound sine plate, the dial indicator will be run across the workpiece twice. The first pass will follow one axis of the part, and the second pass will follow the remaining axis of the part. In this manner, both angles are checked independently.

The Sine Vise and Compound Sine Vise

Although not strictly a measurement instrument, the sine vise and the closely related compound vise are similar in application to the sine plate and compound sine plate. Essentially, these vises are adjustable vises which employ sine plate, protractor, or a combination of both principles to set the position of the vise.

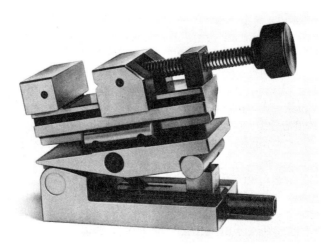

Fig. 5-9. Like the compound sine plate, the compound sine vise is used for setting up workpieces at angles to the base. However, the sine vise is more suited to use in machining operations than the sine plate. (Photo courtesy of Mitutoyo/MTI Corp.)

However, the setting of the angle of the vise is used to determine the angle of the cut in the material. In this respect, the vise is a measurement device. Fig. 5-9 illustrates a typical compound sine vise.

The sine vise pictured here employs the same principles as the compound sine plate, with the angle being set up using a height standard such as gauge blocks. Another style of sine vise uses a Vernier protractor mounted in the base to determine the angle. The compound sine vise of this style uses a second Vernier protractor mounted on top of the angular base to rotate the vise in the second plane. An even more sophisticated configuration of this instrument adds a simple sine plate mechanism under the vise to rotate it in the third and remaining axis. All of these vises are equipped with either holes or slots for attachment to a standard machine table.

Chapter 6

Electronic Measuring Devices

AT THIS POINT in the book, the focus will move toward the new technologies that are capable of finer degrees of measurement, and away from many of the traditional instruments found in the machine shop or toolroom. To one degree or another, manufacturing is being forced to produce products, tools, and machines with greater accuracy and quality. This has largely driven the need and development of instruments with greater accuracy and sophistication. Unfortunately, obtaining the required degree of accuracy has not been possible using traditional mechanical measuring techniques. The development of new electronics such as diodes, liquid crystal displays (LCD), integrated circuits, computers, optics, and lasers has fueled the development of new measurement applications.

Although electricity will play a big role in the devices and instruments discussed in the remainder of this book, this chapter in particular will detail the applications of many of the strictly electronic instruments, leaving optics and lasers for another chapter. The instruments in this chapter will be those which essentially translate mechanical input into electrical signals for display on a terminal. This method of measuration allows for very small distances to be read on a display which would otherwise be

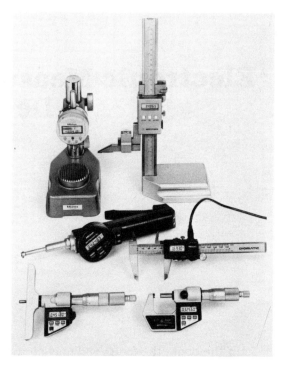

Fig. 6-1. These are a number of different traditional toolroom measurement devices which have been modified to display readings on an LCD screen mounted directly on the instrument. (Photo courtesy of Mitutoyo/MTI Corp.)

impossible to see on a mechanical measuring device, or which could be done faster and more economically with the addition of electronics.

To a certain degree, we have already touched on some instrumentation which is electronic, at least in part. During the discussion of the Vernier caliper, we talked about the electronic display which can replace the standard or dial indicator Vernier scale. This electronic addition is typical of those we will encounter in this chapter, where a mechanical action is translated into electronic impulses and displayed on a terminal, which in this case is an LCD display. Fig. 6-1 illustrates a number of different traditional toolroom measurement devices which have been modified to display their readout on an LCD screen mounted directly on the instrument.

However, electronic measurement is not restricted to electro-mechanical instruments. In practice, any type of energy source can be registered as electric impulses and displayed to a terminal. This includes pneumatic devices which use the flow of air in lieu of mechanical motion, and will, in a later chapter, include optical and laser instruments which translate light signals into electrical impulses.

Electro-Mechanical Measuring Devices

Electro-mechanical measuring devices are those that take motion in the form of linear or angular travel and convert it to electrical impulses. The impulses are then relayed to a second device where they are registered and either displayed or stored. If they are stored, it will most likely be in an electronic format, such as storage in a computer memory or on a magnetic tape or computer disk. These stored data can then later be used in comparison inspection, NC machining, or simply read at a later time by an operator.

Typically, however, the measurement is immediately displayed on another device. This secondary device may be a computer screen, display terminal with diode or LCD character generation, printed on a piece of paper, or a combination of these techniques. Fig. 6-2 illustrates a traditional Vernier caliper with the addition of an LCD readout and a hard copy paper printer to make a permanent record of the measurements being taken. Often, electro-mechanical measurement devices are two separate entities—the measurement device and the readout device. In this way, the method of readout can be chosen independently of the measurement device depending on the requirements and preferences of the individual operator.

The type of display is also largely dependent on the type and sophistication of the measurements being taken. For a coordinate measuring machine, a light diode display featuring three rows of numbers representing the distances along the X-, Y-, and Z-axes may be sufficient; but a television screen or computer screen display (technically called a cathode ray tube, or CRT) would probably be required to interpret all of the information from an operation like laser scan optical microscopy. However,

Fig. 6-2. Traditional Vernier caliper with the addition of an LCD read-out and hard-copy printer for a permanent record of readings. (Photo courtesy Mitutoyo/MTI Corp.)

electro-mechanical measuring devices are generally for simple linear measurements, albeit very small and highly accurate measurements, and will therefore usually have their display in the form of one or more rows of numbers.

For example, in Fig. 6-3, the vertical milling machine has been modified to display the position of the table relative to the spindle on an auxiliary LED (light emitting diode) display. In this case, the machine only displays the X- and Y-axes of travel. The X-axis of travel is the length, controlled by the hand wheel shown in the operator's right hand. The Y-axis is the in and out travel of the table shown here controlled by the hand wheel in the operator's left hand. The motion of the table is registered by sensors and relayed to the display panel above the operator's head. This method is much faster and easier for the operator, versus reading the dial Vernier scales attached to each hand wheel.

SURFACE FINISH AND THE SURFACE INDICATOR

Electro-mechanical measuration instruments are typically used in the measurement of minute distances, such as those essential to fine surface finishes. However, there are several factors included under the concept of surface finish, and an under-

Fig. 6-3. The addition of sensors and digital readout panels increases the ease and efficiency with which the operator can operate a machine. Note the readout on this machine is calibrated to one-thousandth of an inch. (Photo courtesy of Mitutoyo/MTI Corp.)

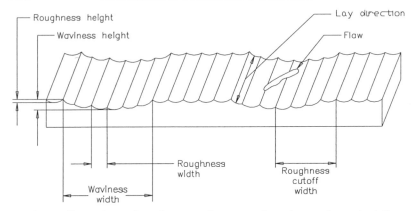

Fig. 6-4. Illustration of surface roughness and waviness characteristics.

standing of these factors is prerequisite to understanding the use of these instruments. We will first cover the factors involved in surface finish, and then the surface indictor and other related instruments. Keep this information in mind, as we will make a short review of these principles in the chapter on optical measuring devices.

Surface finish, in general terms, is the amount of roughness of any given surface, but roughness can take any of a number of forms. Fig. 6-4 illustrates these different forms of surface roughness. This illustration takes into account only a small area of the total surface. In addition to these surface features, the entire surface can have a contour such as run-out, convex, or concave. We will leave these latter considerations until the section on optical interferometry.

Here we are looking for features in the surface's "profile," which is the shape of the surface if we looked directly into the side of it. Profile includes all of the surface features such as roughness height and width, waviness height and width, lay direction of features, flaws, and roughness width cutoff. Depending on the feature being considered, the unit of measure may be either a fractional part of the inch, or directly in microinches which are one-millionth of an inch. In any case, the overall object is to determine how much the surface finish deviates from a perfectly flat plane. This plane is a hypothetical plane called the "nominal profile."

Now we need to define some of these terms, so continue to

refer to Fig. 6-4 in order to have a complete understanding of their meanings.

Roughness is the smallest deviation in the overall surface finish, consisting of little trough-shaped contours in the surface which fall within the distance of the roughness width cutoff. These features are inherently formed by most machining processes during material removal. In other words, these are the lines formed on the surface of a part when the surface is cut with a milling machine, for example. Their height and width are largely dependent on the type of machining operation, the speed of the cutter, and the feed of the part into the cutter.

Roughness height is the average deviation of the surface roughness from a hypothetical line extending the length of the roughness width cutoff. This is *not* the overall height of the surface roughness from the bottom of the trough to the top of the crest. Rather it is the average deviation measured approximately through the center of the roughness contour. See Fig. 6-5. This feature is measured in microinches.

Roughness width is the distance across the crests of the roughness contours. This distance is measured in fractional inches.

Roughness width cutoff is actually not a surface finish contour, but rather an arbitrary distance selected by the instrument operator which will include all of the features involved with surface roughness. Obviously, the roughness width cutoff must be at least as wide as the roughness width, but normally shorter than the waviness width. This distance will be set on the instru-

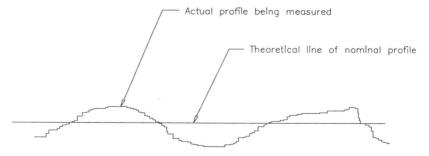

Fig. 6-5. Illustration of roughness height measured in relation to a nominal perfect profile.

ment doing the measuring in order to allow the instrument to do any computational mathematics concerning roughness height. This value is measured in fractional inches.

Waviness looks like roughness, but with a much longer cycle and greater amplitude. In other words, waviness is a longer wave form in the surface contour with roughness contours superimposed onto the waves. This value is measured in fractional inches. Like roughness, waviness is a byproduct of material removal; but rather than being formed by the cutter, waviness is formed by things like chatter in the cutting operation or warpage in heat treating processes.

Waviness height is determined differently than roughness height, in that waviness height is the actual distance from the crest of the wave to the bottom of the trough. This value is measured in fractional inches.

Waviness width is the distance from the crest of one wave form to the crest of the next successive wave form. This value is measured in fractional inches, and again is not so much a matter of the actual surface finish as it is a specified value which is the maximum the design will allow.

Lay direction is another of the byproducts of the machining process. This is the direction of the features of the surface finish as they lay perpendicular to the profile plane. This feature has no measurement value, as it is merely a direction.

Flaws are any of the various surface defects which lay sporadically around on the surface with no real pattern or regularity. They have no lay direction, and generally are not included in surface measurements. These features will generally not affect the surface finish significantly enough to be a concern. However, if they are frequent enough or severe enough, the part may need to be remachined. Flaws are more within the realm of optical measuring devices dealing with surface topography and microscopy and we will consider them in greater detail in that discussion.

Part designs which require surface finishes of a particular degree of smoothness are provided with a symbol indicating to the builder what needs to be accomplished. This symbol, sometimes called a "profilometer symbol" (meaning the values necessary from a profile measurement meter), consists of a square-root symbol with the horizontal extending slightly out over the "V" portion of the symbol. Any one of a series of numbers or addi-

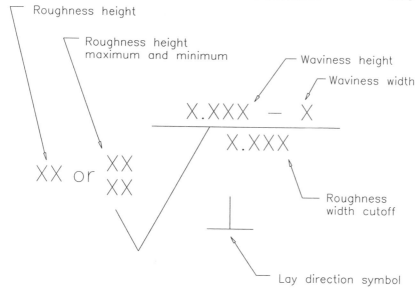

Fig. 6-6. Illustration of the various symbols and numbers associated with the surface profile symbol. (*Note:* A complete listing of additional lay direction symbols and other information about the surface profile symbol can be found in *Machinery's Handbook,* also published by Industrial Press.)

tional symbols can be placed around the surface profile symbol. Reminiscent of a welding symbol, the placement of the numbers around the profile symbol indicates what features they are to determine. Fig. 6-6 is a diagram explaining the meaning of the various numbers and additional symbols associated with the surface profile symbol.

The degree of accuracy and the number of features specified will now determine the type of instrument needed to make the measurement. For now, we will only cover the most basic of these, leaving the more intricate for later. The *surface indicator* is the instrument used to determine the surface profile of a surface that has been machined.

There are several varieties of the surface indicator—including a hand-held type and a table-top or bench-mount type—but they all have essentially the same components for measuring. Fig. 6-7 is a typical hand-held-type surface indicator with an LCD numeric readout. This particular instrument can also be connected to a readout device where the indicator's readings are interpreted

Fig. 6-7. Illustration of a typical hand-held surface indicator. (Photo courtesy of Mitutoyo/MTI Corp.)

and displayed. The display may be either an LCD or CRT display of the numerical values, or a paper hard-copy readout with some combination of the numerical values and a graphic representation of the profile. Fig. 6-8 illustrates a complete surface analyzation system, with the values and graphic representation of the profile shown on a display screen and the values recorded on a paper hard copy.

The spindle at the end of the instrument has a small, and extremely sharp, diamond stylus that is brought down to contact the part's surface. The instrument then moves the stylus over a short distance of the surface allowing the point to follow the surface contour. The motion of the stylus is translated into electrical impulses which are read by the instrument. Inside the surface indicator, these impulses are registered and the results are amplified to some even factor when they are displayed.

For example, if the part surface is comparatively smooth, then the amplification factor might be 10,000. This will produce a graphic representation of the readout that is easily defined. Conversely, if the part surface is obviously rough, then the amplification factor could be set on the instrument by the operator to perhaps 20. Again, this will allow for a readout that is easy for the device to display and for the operator to read.

The surface indicator is really nothing more than a "transducer" (which we will cover shortly) attached to an amplifier. The instrument translates the roughness and waviness patterns of a

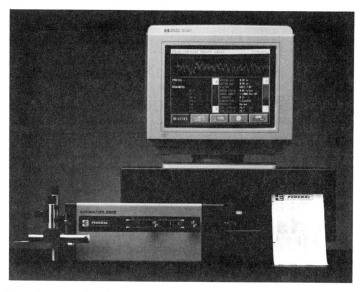

Fig. 6-8. The surface analyzer can be attached to various readout devices for greater interpretation of collected data and permanent records. (Photo courtesy of Federal Products Co.)

surface into a larger scale. However, the electronics of this instrument on some models will also give the operator a mathematical average value of each of the surface roughness conditions discussed earlier. These averages then can be used to determine if the part meets engineering surface finish requirements.

The surface indicator is also applied in another form in an instrument called a "roundness measuring machine." This instrument is used for measuring the surface contours of internal and external diameter. The workpiece is placed exactly in the center of a rotary table, and the probe point of the surface indicator is brought to bear against the surface. The machine is then zeroed and the part rotated, allowing the surface indicator probe to follow along the surface. The result is the same as the linear surface indicator. Some roundness measuring machines are also equipped with a rotary hard copy paper recorder which draws an enlarged representation of the circular profile.

TRANSDUCERS

The "transducer" is literally a device that translates linear motion into electronic impulses, which in turn are displayed as numerical values either on a terminal screen or a computer. It is a comparative device, in that it compares the amount of travel of the spindle to a given standard. For example, the transducer can be used in the same manner as the dial indicator, where the device is brought into contact with a surface, zeroed, and then traversed along a surface, thus registering variations in the contour or attitude of the surface.

The transducer consists of a housing, typically with a machined shoulder on one end, and a spindle extending from the shoulder. The shoulder is machined to an exact tolerance and used as the mounting area for the transducer. Typically, this shoulder is placed into a bushing located at an exact position relative to the part being measured. This application is common in applications of checking fixtures where the fixture holds the part in exact position relative to the bushing, and the transducer measures any variation in the part's size or configuration.

Inside the housing is the mechanism of the transducer for converting the linear motion of the spindle into electronic impulses. Again, the description vaguely resembles that of the dial indicator. However, unlike the rack-and-pinion mechanism of the dial indicator, the mechanism of the transducer is typically entirely electronic. The advantages of this are durability, accuracy, and of course the opportunity to feed readings directly into a computer, CNC machining center, or other electronic device.

Although similar in principle to the surface indicator, the transducer will not usually involve an amplification device. The amount of travel of the spindle is directly proportional to the amount registered on the display. While the surface indicator will take surface readings in microns, the transducer is typically restricted to one ten-thousandth of an inch (occasionally to one one-hundred-thousandth of an inch).

ELECTRONIC COMPARATORS

A specialized application of the transducer is an instrument referred to as the "electronic comparator." The electronic comparator places a transducer probe on the end of an arm-and-stand combination resembling a combination of a dial indicator mounted on a height gauge. Using a surface plate and a linear standard such as a gauge block or other standard, the electronic comparator is brought to zero, and then the probe point is moved to the top of the workpiece being measured. This entire process is exactly the same as for a dial indicator, but is somewhat more accurate.

If the exact linear dimension of the workpiece in question is not important, but rather only the tolerance of the surface, then the electronic comparator can employ a light indicator. Once the instrument is zeroed, the probe is passed over the top of the workpiece surface and a light will come on if the part is either above or below the specified tolerance of the part. Some commercially available electronic comparators will employ two different colors of lights to indicate whether the part is above or below the specified tolerance. Obviously, the tolerance parameters of the part are set on the instrument prior to checking it.

THE COORDINATE MEASURING MACHINE

The Coordinate Measuring Machine (CMM) is a highly sophisticated type of robot with a tiny probe transducer for a hand. The transducer registers input from the probe as it contacts a part, and sends that input as electrical impulses to a computer. The computer in turn interprets the impulses and records them as specific points in space. See Fig. 6-9.

Here it helps to understand a little analytical geometry. Every point in space has a location in three axes: up/down, in/out, and fore/aft. The method of charting these positions was conceived by the seventeenth century philosopher and mathematician René Descartes (day-kärt'), therefore it is called the Cartesian (kär tEE' zhuhn) Coordinate System. Fig. 6-10 illustrates the three-dimensional Cartesian coordinate system.

Points in space from left to right (fore/aft) are recorded along the horizontal axis, called the X-axis. Points in space from top to bottom (up/down) are recorded along the vertical axis, called the Y-axis. Points toward or away from the observer (in/out) are recorded along the Z-axis. Since the Z-axis travels theoretically directly away from the observer, it is traditionally shown rotated at 45° for clarity.

The designations X, Y, and Z for the axes are also universally recognized as a matter of tradition. The point where all of the axes cross will always be zero. Positive numbers are to the right on X,

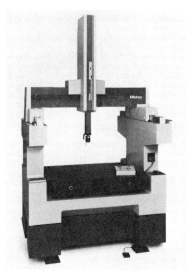

Fig. 6-9. The Coordinate Measuring Machine (CMM). (Photo courtesy of Mitutoyo/MTI Corp.)

up on *Y*, and toward you on *Z*. Negative numbers are obviously recorded on those same axes in the opposite directions.

Fig. 6-11 illustrates how a point in space is charted on the Cartesian coordinate system. The numbering system for charting these positions is always noted in the same order which, as one would expect, is alphabetically. The *X* coordinate will always be

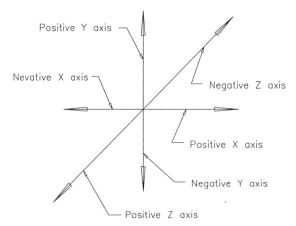

Fig. 6-10. Illustration of the three-dimensional Cartesian Coordinate System.

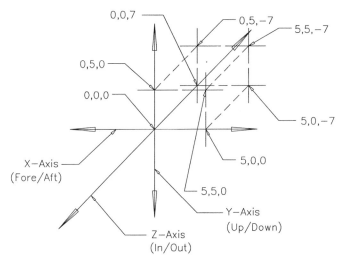

Fig. 6-11. This illustrates how a point is located in three-dimensional space on the Cartesian Coordinate System. This is a typical application of "analytical geometry."

listed first, followed by the Y coordinate, and finally the Z coordinate. In the illustration, the point being located is at $X=5$, $Y=5$, and $Z=-7$. If the CMM is shown the location of this point, relative to the 0,0,0 origin, it will know exactly where to look for each of the other points shown.

The CMM fixture is designed to hold the workpiece in the exact position it will be in during actual use on whatever assembly it is being made for. In this way, every point on the part will be in its exact position in space, relative to every other point on the part. A known point in space is located, usually on the fixture but occasionally on the part being checked, and the coordinate measuring machine is located relative to that point. It then begins probing the remaining points on the workpiece to determine their exact location. (See Fig. 6-12.)

All of this sophisticated fixturing and computer-aided robotic rigmarole may seem like a bit much to the inexperienced individual—just to see if a part is the right size and shape. However, checking complex compound warped surfaces would be incredibly difficult or impossible without it. This type of checking is especially suited to checking automobile body panels, aircraft and spacecraft body parts, ship screws (propellers), airplane propellers, and jet engine blades. In simplest terms, if you are called

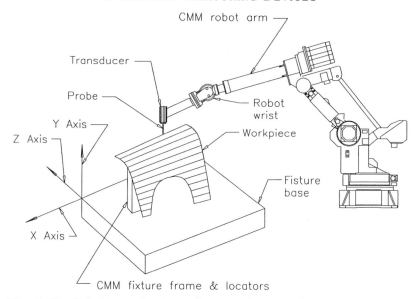

Fig. 6-12. Robotic application of a transducer probe in a quasi-CMM application. Note the fixture used for holding the complex shape in an exact and consistent position. (Robot graphic courtesy of Fanuc Robotics.)

upon to check something shaped like a potato chip, CMM checking is probably your best bet for getting all of the correct measurements.

ELECTRONIC LEVELING DEVICE

Like many of the more sophisticated measuring devices, the electronic leveling device requires a thorough study of the manufacturer's documentation because each device will vary slightly from one manufacturer to the next. However, the principle on which the device operates is easily understandable, and it is advisable for the student of measuration to be familiar with the general principle of this instrument.

As the name suggests, the electronic leveling device is an instrument used to determine if a plane is parallel to the earth's gravitational pull. This is a concern for such operations as setting a surface plate, setting a large machine, or setting a bolster plate within a large machine or press. Fig. 6-13 illustrates a typical electronic level.

The electronic level consists of a system comprised of one or

Fig. 6-13. Illustration of a typical electronic level system. (Photo courtesy of Federal Products Corp.)

two "leveling heads," a "base" for each head, a set of connector cables, a readout device, and (optionally) a computer. Each leveling head sends a signal via the connector cable to the readout device indicating the attitude that the head is sitting relative to the force of gravity. When an optional computer terminal is utilized, it can use these data with specialized software to make determinations about the part being checked.

The leveling head consists of a container, about the size of a normal textbook, attached to an adjustable base. Inside the leveling head (illustrated in Fig. 6-14), a pendulum is suspended from a strap, or "read spring," at each end of the pendulum. Above the pendulum is an electric induction coil, or "core," and a loop of wire, called the "shading loop," attached to the pendulum which extends up and around the core. As the leveling head tips to set on a surface, the pendulum will swing down to remain toward the earth. This in turn shifts the position of the shading loop and causes an imbalance in the electric flux of the core, which sends a signal to the readout device where it is registered and displayed. The resulting angular measurement is extremely accurate, and generally measured in seconds of arc.

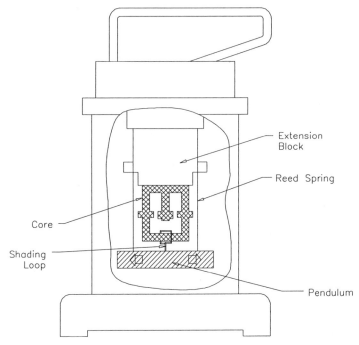

Fig. 6-14.   Illustration of the internal mechanism of the electronic level.

Electro-Pneumatic Measuring Devices

Electronic measurement instruments, other than those involved in measuring electric current and its associated properties, typically employ some secondary means of making the measurement. In the previous section, that means took the form of mechanical motion. But some electronic measurement instruments are able to take advantage of the motion of air. Air-operated instruments, whether they are operating a wrench or a measurement instrument, are referred to as "pneumatic," which is derived from the Latin and Greek words for "air" or "to breath."

The electro-pneumatic measuring device utilizes an air flow and a means of registering the amount of resistance to that air flow. Therefore, electro-pneumatic measurements are generally used for taking internal measurements where the flow of air from the instrument can be either partially or completely trapped. However, external dimensions can also be measured, provided the part can be trapped inside of the air flow, such as a pneumatic ring gauge. One of the most common applications of this technique is used for measuring the inside diameter of holes. Obviously, due to the very nature of the device, the workpiece being measured must be of a material which is nonporous, because a porous material would allow excess air to escape from the cavity and produce an inaccurate measurement.

Obviously, there must be some reason for using this type of instrument for measuring the diameter of a hole, when an inside micrometer or plug gauge could just as easily be employed. Industry has for some time now been looking for more ways to do "nondestructive" or even "noncontact" testing and measurement. Some part surfaces are machined to such high surface finishes and such close tolerances that even contacting them with a checking instrument, such as a plug gauge, can mar the surface or push a feature out of tolerance. Another problem is that some materials are very soft, and contacting them can mar the surface easily. The problem for industry therefore became how to measure a part without actually touching it. Part of the answer came with the advent of light source measuring instruments, which we will cover in the next chapter. However, before the advent of many of the light source measuring instruments, air was employed to contact the surface in lieu of a physical contact with a hard measuring device.

There are predominantly two types of electro-pneumatic measuring devices used in industry: the "column type" (or "flow type" as it is sometimes called), and the "pressure type." The applications of these two types of electro-pneumatic instruments are similar.

COLUMN-TYPE PNEUMATIC GAUGE

The column-type electro-pneumatic gauge operates on the "velocity" at which the air flows. This device consists of an air filter, air regulator, a vertical flow column, an indicator float, a flow control valve, and a gauging head. Fig. 6-15 illustrates the arrangement of components in the column-type pneumatic gauge.

Air is supplied to the instrument from a compressed air system typically found in most manufacturing facilities. However, this air is normally maintained at pressures ranging from 40 psi

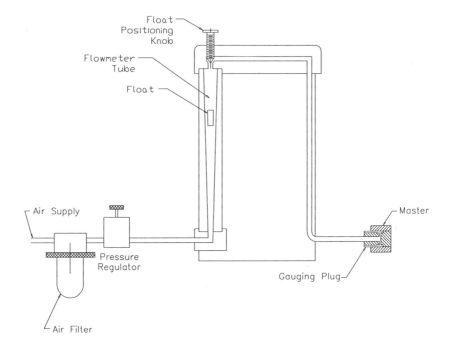

Fig. 6-15.   Illustration of the mechanism of the column-type pneumatic gauge.

(pounds per square inch) up to 250 psi. The pressure required by the electro-pneumatic gauge is about 10 psi. This is accomplished by placing a flow control valve at the beginning of the instrument setup. Air pressure is reduced to the appropriate amount before ever entering the measuring system.

Next, the air delivered to the device, and consequentially to the measuring process, must be completely free from any airborne contaminants such as dust particles or grease. The air is therefore passed though a canister-type air filter to remove any impurities. The canister hangs below the air line to trap moisture as well as particles. Periodically, the regulator valve is closed and the filter canister opened, and the filter material is replaced to assure proper operation of the instrument. This filter is purposely placed after the regulator valve, not only as a means of shutting off the air flow for filter replacement, but to maintain proper operation of the filter. If the filter were placed before the regulator valve, the high pressure of the shop air could force unwanted particles through the filter material.

The filtered air then enters the bottom of the indicator column. This column consists of a clear tube (generally plastic) that is tapered from top to bottom. This allows the float to be suspended in the column, riding on the flow of air through the column. Since the tube is tapered with the large diameter of the taper at the top, it will take continually greater flows of air to raise the float higher in the column. The float itself is a sort of little flanged parachute that rides on the column of air, and its position is registered on a scale on the air column. At the top of the column is a float positioning control valve which positions the float in the column by restricting the air flow out of the column, while the instrument is being calibrated.

Finally, the air flows through a flexible plastic tube and out through the gauging head. The gauging head consists of a tubular handle for the operator to grasp, and an interchangeable gauging head. The actual head may be either a ring- or a plug-type gauge head, depending on the application (although plug gauging is more common).

The instrument is calibrated by using a master gauge. For example, using a plug gauge style gauging head, the master gauge will be a ring gauge of the exact diameter to be measured. The plug is placed in the master gauge with the air flowing, and the float positioning valve is adjusted until the float is hanging at the zero line of the scale along the column. An indicator is then set at the upper and lower limits of the feature being measured.

Measuring a hole in this manner simply now involves placing the plug gauge into the hole and reading the amount of variation on the column scale. If the hole is oversized, excessive air will be allowed to flow from the gauging head (i.e., the air will escape from the head more quickly—remember this instrument actually measures the velocity of air flow) and the float will therefore rise higher in the column. If the hole is undersized the air flow will be restricted and will therefore escape more slowly, and the float will fall to a lower level in the column. Depending on the particular instrument and the calibration, this type of measuration device is capable of amplifying the reading of the hole from 1000:1 up to 40,000:1.

Pressure-Type Pneumatic Gauge

The pressure-type pneumatic gauge has essentially the same types of applications as the column-type pneumatic gauge; the only real differences are a slightly lower degree of magnification of the pressure type, and a physically different mechanism of registering measurements. Fig. 6-16 is an illustration of the system used in the pressure-type pneumatic gauge. Note that the pressure-type pneumatic gauge uses standard "shop air" supply just like the column type, which is filtered and regulated before entering the gauge.

Once the air flow enters the gauge, it is split into two separate, and more or less equal, streams of air. One air flow is allowed to escape to the open atmosphere from the "reference channel" and the flow rate is regulated by a zero setting valve. The remaining stream of air is channeled to the gauge head via the "measuring channel." The measuring air flow then escapes to the atmosphere through jets in the gauge head. This gauge head performs the measurements in a manner similar to that of the column-type pneumatic gauge, by measuring the amount of air that is able to escape from the gauge head during the measuring process.

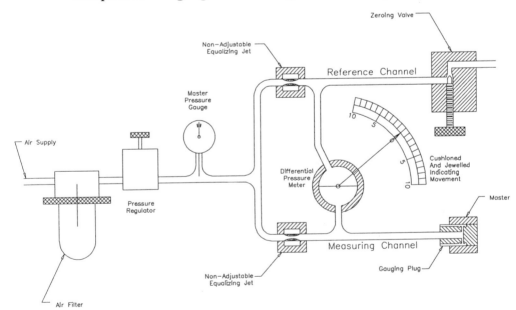

Fig. 6-16. Illustration of the mechanism of the pressure-type pneumatic gauge.

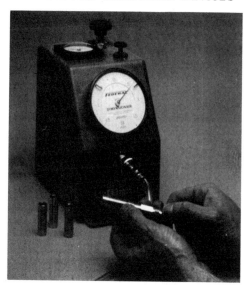

Fig. 6-17. Application of a typical pressure-type pneumatic gauge using a plug gauge head. (Photo courtesy of Federal Products Corp.)

The reference channel and the measuring channel are connected by a final channel with a highly sensitive pressure measuring indicator in the middle. In order to set the pressure-type pneumatic gauge, the gauge head is placed in a gauge "master" of precisely the size of the feature to be measured. With the gauge head in the master, the indicator is brought to zero by adjusting the flow (pressure) which escapes from the reference channel with the zero setting valve. Once this is completed, measuring can begin.

Fig. 6-17 illustrates the pressure-type pneumatic gauge. In application, if the feature being measured were, for example, a hole, a too-large hole would allow extra air to escape from the gauge head, thus dropping the pressure on that channel, and the indicator would fall in that direction. Conversely, if the hole were too small, this would additionally restrict the air flow and increase the pressure in the measuring channel, thus forcing the indicator toward the reference channel. Simply put, the entire system of this measuring device depends on both streams of air remaining exactly equal in pressure.

Chapter 7

Optical Measuring Devices

AT THE BEGINNING OF THE BOOK, we mentioned that the international inch and meter are based on a certain number of wavelengths of monochromatic light; this method was chosen because, for the most part, light is very stable and consistent in the length of its waves. There are a number of other reasons why light-measuring instruments are so vital to industry. We, as a species, rely heavily on sight for many of the measurements we take in industry; therefore, it was a logical jump to start using visual comparison instruments like the microscope as a measuration instrument. This is especially true in an age when miniaturization is abounding, not only in the field of integrated circuit technology, but in fields like biomedical engineering.

Another of the advantages of optical measuring instruments which rely on sight is that the line of sight is absolutely straight. There is absolutely no sag, bend, stretch, or motion in a line of sight. This principle is used in alignment devices such as the jig transit or Theodolite measuring systems. These devices essentially use a telescope to project a line between the instrument and a reference point. In industry, this technique can be used for operations such as setting, aligning, and leveling large machines or fixtures.

Optical measuring devices are also able to take consistent

high-speed measurements constantly without actually touching a part. This allows for high speed mass production to continue without the need to take parts off the assembly line and sporadically check them. This application will be discussed in greater detail when we get to the section on the laser micrometer.

Finally, optical measurement instruments allow us to take measurements that are simply too small to take any other way. The surface finish on the end of a gauge block, for example, is measured in millionths of an inch. This would be virtually impossible without the aid of the optical flat and lightwave technology. Squareness, distance, flatness, and other measurements previously done with conventional tools are now increasingly becoming the domain of optical measuring devices like the optical comparator, microscope, and the laser.

Optical Comparators

In 1921, James Hartness came to the realization that measuring very small objects, especially those with complex contours, would be a lot easier by projecting an image of them onto a screen at a greatly increased scale. The instrument he invented to make these sort of shadowgraph measurements was the optical comparator. As one might guess, this instrument creates an enlarged image of a workpiece on a screen, and uses that enlarged image to compare to a standard. Fig. 7-1 shows the typical application of an optical comparator for examining and measuring small parts. This particular machine has a 30-inch-diameter display screen.

The typical optical comparator consists of a light source, a workpiece fixture or "stage," a series of mirrors to direct the light toward a screen, and finally the projection screen. The light beam travels toward the screen and is broken by the contour of the workpiece, and the resulting shadow appears on the back of the projection screen. Obviously, the position of the workpiece in the light's path is held very closely in order to maintain an exact amount of magnification.

On the face of the projection screen can be attached any one of countless translucent comparison charts. These comparison charts are made of either frosted glass or matte finish plastic, and

Fig. 7-1. The optical comparator displays an enlarged shadowgraph of the workpiece which is then measured. (Photo courtesy of J & L Metrology.)

are etched on the frosted side with a highly accurate measurement system or predetermined contour. These charts are very accurate for an obvious reason. Generally made to an accuracy of 0.0002″, the chart can therefore determine a part accuracy equal in percentage to the magnification of the machine. In other words, if the chart is accurate to 0.0002″ and the magnification of the comparator is 10, then the accuracy with which the chart is capable of measuring the part is 0.00002″. Likewise, if the comparator has a magnification of 50×, then the chart will measure the part to an accuracy of 0.000004″.

The optical comparator is perhaps the simplest of the optical measuring devices, although it operates on a principle similar to that of the measuring microscope. It is best adapted to measuring those workpieces which would be too small and too complex to measure by any other means. Small but highly accurate screw threads, gear teeth, or other similar complex shapes are good candidates for measurement on the optical comparator.

The workpiece is held in the optical comparator by a small fixture, clamp, vise, or jaw chuck in what is called the "staging area," or simply "stage." The stage itself can then be mounted to any one of a number of mechanisms to allow for positioning and movement of the stage. Typically, the stage is mounted on a set of slides actuated by small lead screws to allow for lateral movement, and a small dial table to permit rotary or angular motion. The clamping mechanism of the stage holds the part securely in position, while the slides and dial allow the stage, and therefore the workpiece, to be properly positioned so that the shadow of the part falls in the correct location on the comparison chart.

An alternate type of optical comparator uses a reflected image of the workpiece rather than a shadow of it. The instrument operates the same in this case, but the light source is located above the workpiece instead of below, and is reflected back through the mirrors to the screen. This procedure works nicely, but requires a higher intensity light source to produce the image on the screen. The direct image optical comparator gives the operator a better look at the detail and contour of the workpiece.

The optical comparator finds great application in measuring and checking objects which are too large for the measuring microscope, but have features too small for regular visual inspection. For example, the typical printed circuit board used in a computer might be too large to view with the measuring microscope. However, the staging area of the optical comparator is much larger and able to view this type of component. The copper circuits of the printed circuit board can then be inspected and measured to check if they meet engineering standard for size. This operation would lend itself best to the reflected light type of optical comparator.

The shadow projection style of optical comparator is well suited to inspection and measurement of small, fine-detailed components such as microgears, watch gears, internal winding plates of small servo motors, or small pin connectors used in large multicircuit electrical connectors. For example, the electrical connectors on the underside of a military airplane's wings, used for holding extra fuel tanks or ordnance, have as many as several hundred individual pin connections inside of the connector housing. Each pin and receiver is an individual component that must meet exact standards of size and shape. Likewise, the

Fig. 7-2. The principle of the optical comparator can be adapted to specialized measuring instruments such as this machine tool projector. (Photo courtesy of Stocker & Yale, Inc.)

retainer that holds these pins contains a series of many very tiny holes located in an exact pattern, the position of which can be checked with an optical comparator.

There are also special applications of the optical comparator principle. For example, in Fig. 7-2, a specialized reflecting light style optical comparator is mounted directly to a machine (a lathe, in this case) to measure and inspect the tool bit in place. This instrument is called a "machine tool projector." In this way, the tool holder of the machine becomes the stage, and the comparator can measure not only the tool, but its position in the holder. This same instrument can also be used to inspect and measure the workpiece in the machine.

Measuring Microscopes

Closely related to the principle of the optical comparator is the "measuring microscope." The measuring microscope, sometimes called a "toolmaker's microscope," differs slightly from most common laboratory microscopes in that the light source is provided from "above" the object being viewed rather than from below. It also differs because the viewing stage is actuated by a series of micrometer dials to allow for the precision measurement. Although this is gradually giving way to digital readout, the precision lead screws to actuate the staging table still remain. Fig. 7-3 illustrates the measuring or toolmaker's microscope. Note that this particular instrument is also using a camera mounted at the top and a CRT display screen at left. The addition of these auxiliary devices allows the measuring microscope to more closely inspect the workpiece while measuring features. It also provides a means of having more than one person observe the workpiece at a time. Permanent records of this inspection can also be stored on VHS video tape for later review or reference by a customer.

The optical field of view of the measuring microscope incorpo-

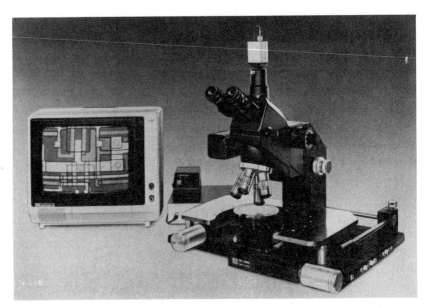

Fig. 7-3. The measuring or toolmaker's microscope can be used in conjunction with a camera and display screen or tape recorder. (Photo courtesy of Mitutoyo/MTI Corp.)

rates a set of crosshairs (reticle) for determining reference points on the workpiece being measured. Once a particular reference is sighted, the micrometer screw handles which actuate the motion of the stage, or the digital readout, are zeroed and the part moved to a point where the crosshairs sight onto the next point of reference. Once this is completed, the measurement from the first reference point to the second reference point is read directly from either the micrometer actuators or the digital readout.

There are other methods by which measurements are taken with the measuring microscope, and other styles of mechanism, which the operator will need to study from the individual instrument's manual, but all will rely on visual acuity and motion of either the workpiece or the field of view of the microscope. However, unlike the optical comparator, the measuring microscope will give the operator a much better opportunity to visually inspect the workpiece (even better than the direct imaging optical comparator). The very nature of this instrument obligates its use to very small parts; and observing defects such as flaws, cracks, or misshapen contours would be virtually impossible without some degree of magnification of the workpiece. In other words, the measuring microscope allows the operator to make judgements about the part beyond the realm of simple measurement. The measuring microscope should not, however, be confused with the "stereoscope" which looks very similar, but which has a fixed stage and is typically used only for visual inspection and not measurement.

There are many optional pieces of hardware that can be used with the measuring microscope. For the measurement, there are optional types of digital readouts which mount either directly to the actuator handles or on a separate readout stand. The reticles are interchangeable in much the same manner as the comparison charts on the optical comparator to utilize various configurations of crosshairs or graduations. The eyepiece itself is interchangeable with a protractor eyepiece, a binocular eyepiece, a rotary template eyepiece, a double-image eyepiece, and the like. Special cameras can be attached to the instrument for taking photographs of the workpiece under magnification. Various hard-copy readout devices can be attached to the measuring microscope for permanent record of the measurements taken. Finally, there are various fixtures for the stage to carry virtually any shape part and several configurations of light sources (with additional various filters) to accommodate proper lighting of any shape of part.

Optical Leveling and Angularity Devices

Perhaps the most common form of optical measurement is that used for angularity over comparatively long distances. This includes angles toward the horizon (levelness), perpendicular from the horizon (elevation or plumbness), and parallel to the horizon (azimuth or bearing). While generally thought of as being measurements used in surveying and military tactics, these forms of measurement are actually common in architecture, marine vessel construction, and building and placing large industrial machines.

Optical angularity instruments operate on a fairly simple principle, although there are some rather complicated calculations involved in many measurements. The basic principle of the optical angularity instrument is that light, and therefore the line of sight, travels in an absolutely straight line for an infinite distance, and with absolutely no size to the line. For example, if a builder were trying to lay out a grade for a construction sight using a string stretched between two points, the line would always sag to some degree because of the weight of the string, no matter how taut the string were pulled. Additionally, the string itself has some size, regardless how small. If the builder were trying to determine a level grade, the situation would be exacerbated by having to suspend a small level from the string and thus increasing the sag of the line.

The line of sight, however, has no weight for gravity to act upon, and no size to interfere with measurements. It merely travels from point A to point B with no deviation. Determining the angle of this line of sight now becomes the function of the instrument used to create the line. In the following sections, we will cover some of the optical instruments used to determine and analyze the line of sight used in angular measurements.

TRANSIT

The transit utilizes a low power telescope mounted in a frame, which in turn is mounted on a tripod stand, to create a line of

Fig. 7-4. The transit uses a series of leveling devices in conjunction with a line of sight telescope for alignment, leveling, and squareness measurements. (Photo courtesy of The L.S. Starrett Co.)

sight between points. Once a line of sight is established, the attitude of that line can be read directly off the instrument, and consequently can be compared to the reading for successive lines. The frame which holds the telescope has several features: a leveling device, a horizontal protractor, a vertical protractor, and (optionally) a laser or infrared range finder. Suspended under the frame of the transit is a plumb bob for setting the location of the instrument directly over a specific point.

The telescope itself consists of an objective lens at the front end, a focusing lens and mechanism, a reticle, and an eyepiece. Fig. 7-4 illustrates a typical transit. The objective lens and the reticle are stationary on either end of the telescope, with the focusing lens and eyepiece adjustable along the axis of the scope's tube. When the image sighted with the telescope passes through the focusing lens, it is projected upside down on the reticle. The lens in the eyepiece both inverts, or uprights, the image and magnifies both the image and the reticle pattern. This allows the operator to observe the reticle pattern projected directly on the point where the line of sight strikes the object being sighted.

Setting the line of sight and the point on the object being sighted requires some practice. Due to the very nature of the human eye, sighting of a transit can cause a condition called "parallax." Parallax is the appearance of a difference in position caused by looking at a point from two or more different points in space. This means that if the operator sights a point through the transit and then moves the eye in location on the eyepiece, the pattern of the reticle will appear to land on a different place on the object being sighted. This is partially why both the focusing lens "between" the objective lens and the reticle is movable and the eyepiece is also movable. The focusing lens produces a sharp image on the "screen" of the stationary reticle, while the eyepiece focuses the shorter distance from the reticle to the operator's "individual" eye. I stress "individual" eye here because the human eye varies, and the eyepiece will adjust the necessary "diopter" for each individual operator.

The first order of business when using a transit is setting up the instrument so it will function properly. Initially, the tripod is set up so it is stable, and positioned so the plumb bob is directly over a specific starting point. Next, the frame holding the telescope is leveled. The frame of the telescope is mounted to the tripod on four small leveling screws. Once the tripod is set, the telescope is leveled first in one direction, and then at 90° to that direction by adjusting the leveling screws. The process is repeated from one axis to the other until the transit remains level regardless of how the frame is rotated. This provides the instrument with a starting plane that is perfectly parallel to the horizon.

Leveling the transit telescope requires another device on the telescope frame to determine the attitude of the frame. We have referred to theoretical planes which are parallel or vertical with respect to the horizon. These planes are actually lines of sight in relation to the pull of gravity at the placement of the transit. It is assumed that the pull of gravity is exactly perpendicular to the horizon at the place where the transit sits. Therefore, setting the transit's line of sight parallel to the horizon actually means that the telescope is sitting perfectly "level." Level, by definition, means parallel to the horizon.

Attached to the frame of the transit are a number of "levels" which operate just like a normal manual level or the common carpenter's level. There are several forms of these devices, and the type and number of them will vary from one instrument and manufacturer to the next. The "spirit level," "coincidence level," and "plate level" are some of the most common types of levels found on the transit. They will, for the most part, all operate on a similar principle, but for our purposes we'll describe only the spirit level.

This level consists of a glass tube formed in a curve and fixed into a holder which seals the tube. Within the tube is a low viscosity fluid, typically lightly colored, with sufficient space left to form a bubble in the fluid. Typically this fluid is an alcohol or other chemical with a low freezing point so the instrument can be used in any temperature environment. The tube is etched with graduations at either side of the center of the top of the arc of the tube. Since the bubble will automatically rise in the fluid directly away from the pull of gravity, it will always be as high in the tube as possible. As the adjusting screws of the transit frame are moved, the bubble will travel along the tube. When the bubble is exactly equidistant from one end to the other from the center graduation of the tube, the frame is level.

While other types of transit levels work slightly differently, nearly all will employ a fluid-filled tube with a bubble and a graduated glass. It is important to read the manufacturer's manual for each instrument to be certain of proper operation of the particular type of level employed on that particular instrument.

Determining a measurement by line of sight requires two components: an instrument to sight "from," and a target to sight "to." The transit serves as the instrument to sight from, and the target can be any one of a number of devices. Typically, in surveying, the sight target is a graduated stick called a "scale," or in surveying jargon a "story pole." When the transit is used to align buildings or machines, the sight target may be a construction ball, a specialized optical target, or even a particular feature of the object being measured.

Sighting a transit for industrial applications, such as leveling a fixture or machine, is somewhat more involved than for surveying. Industrial transit use frequently involves specialized targets to fit a particular application. Often the target will contain a set of crosshairs similar to those of the reticle of the transit, and the two sets of graduations must be aligned during sighting. Another instrument used in conjunction with the transit is a "collimator" or "auto-collimating telescope." Yet another type of target is an enclosure with a mirror set in it at a 45° angle to allow for sighting around a corner to a still further target. This will establish two perpendicular lines of sight more or less simultaneously: one from the transit to the corner target, and a second from the corner target to the final target.

Collimation, auto-collimation, and specialized targets would fill a volume entirely on their own, and therefore fall outside the purview of this book. The application of the transit is an exercise in basic trigonometry. Once the point at which the transit sits is established, a second point for the target is established. Typically this happens by starting with the transit at a "benchmark." The distance to the second point is then measured to a linear distance and established (relative to magnetic north from the benchmark, in the case of surveying).

The elevation of the first target sighted to will undoubtedly be measured with a scale. The transit is sighted to the first target and the protractors set to zero. We now have enough information to begin sighting the locations of remaining targets, i.e., two points and a line of given length. Making the readings and calculations necessary for using a transit comprise an entire college course by itself. This section, even though it is lengthy, is only a glimpse at understanding the transit, and it is recommended that a specific text on transit operation be consulted before trying to operate this instrument.

Lightwave Precision Measurement

Lightwave precision measurement takes one of several forms, ranging from the use of ambient light to the use of laser beams. It also can be concerned with any one of a number of various measurements. In addition to linear measurements and angularity, precision lightwave measurement is used to determine straightness, flatness, plumbness, squareness, runout, parallelism, roundness, concentricity, and other conditions, most of which deal with surface finish, shape, or position.

All optical systems operate to one degree or another with the assistance of light. The transit, for example, uses ambient light to reflect the image of the target to the operator's eye. More sophisticated systems might use laser light, which we will discuss in the next section, to produce certain measurements. Light, however, is only a small fraction of what we call the "electromagnetic spectrum." The electromagnetic spectrum includes all of the visible light, and a number of things above and below visible light. Radio and television waves make up the lower end of the electromagnetic spectrum, and X-rays and Gamma radiation make up the higher end.

In measurement, we will only be concerned with the area of the spectrum from just below the visible light range to just slightly above it. The area just above the visible light range is called the "ultraviolet" spectrum, and this is, for measuration purposes, the domain of lasers only. The visible light range is next lowest, and is used by nearly all optical measuration instruments. Next lowest on the spectrum is the "infrared" spectrum. Finally, at the lowest range of measurement concern is the "microwave" spectrum. This is again the domain of lasers only, and in the most technical sense this becomes a specialized laser called a "maser."

As we progress into the discussions of optical flats and interferometry, it will be necessary to have some understanding of topography. Topography is normally thought of in terms of mapping contours of the earth over a specific area. However, topography is actually the mapping of contours over any area, no matter how small, showing all of the elevations, depressions, deformations, and shapes. By using lines which represent progressive elevations from a benchmark, the topographical map can interpret three-dimensional features in a two-dimensional plane—in this case, a map.

The lines on a topographical map tell the reader outright what the elevation is, and therefore how it is different from the elevations that preceded it—either higher or lower—and by how much. Obviously this is not the case in lightwave precision measurement; and reading the topography of a workpiece becomes a matter of learning what the patterns of lines are telling you.

Optical Flats

The optical flat is a light-operated instrument used to determine surface runout or contour of a workpiece. An optical flat doesn't look much like what you might expect a measurement instrument to look like. It consists of a round, flat disk of highly polished, compressed powdered quartz, about three inches in diameter and one-half-inch thick. The flat surfaces of the disk are lapped to within a few millionths of an inch of perfect flatness.

Fig. 7-5. The optical flat utilizes light interferometry to measure size and runout of flat surfaces. (Photo courtesy of The DoAll Co., Des Plaines, IL.)

The material itself is completely transparent. Fig. 7-5 illustrates an optical flat.

Obviously this type of an instrument cannot be used without some auxiliary equipment, in this case a light source. While the optical flat will produce the contour lines indicative of measurement using ambient light, there is no way of telling the value of these lines. The reason is that the wavelengths of ambient light vary dramatically and are also of mixed color, and therefore of mixed wavelength. To alleviate this problem and produce an accurate measurement, the optical flat is used with a specialized light source using a single wavelength. Commonly this light is produced from passing white light through helium, which produces a greenish-yellow light with a wavelength of 23.1323 millionths of an inch.

The optical flat is used to determine surface flatness. The two flat sides of the disk are somewhat different, and therefore the top surface is indicated by an arrow on the side of the disk. While the disk is transparent, the lower surface is both transparent and reflective. This causes all lightwaves that strike this surface to be split into two parts. One part is reflected back to the top, and the other part is allowed to pass through the lower surface and proceed to strike the surface of the workpiece.

The optical flat is placed directly on the surface of the workpiece to be measured. Depending of the application of the optical flat, this workpiece surface may be the entire support for the instrument, or part of the instrument may be supported by a gauge standard such as a gauge block. Once it is placed on the surface of the workpiece and subjected to the helium light, dark bands will appear on the workpiece surface. Each of these bands represents a certain amount of runout of the part. The distance between the bands is essentially immaterial, only the number of bands is important. Even though the greater the amount of runout, the greater the number of lines there will be, and therefore they will appear closer together, the exact distance between them means nothing.

The dark bands formed on the surface of the part during measurement with the optical flat are actually places where the reflected lightwaves from the lower surface of the optical flat interfere with the lightwaves reflected from the surface of the part. Fig. 7-6 illustrates this interference condition. This interference of lightwaves causes a cancellation of the light at the point of intersection, resulting in the appearance of the dark bands.

Since each wavelength of helium light is 23.1323 millionths

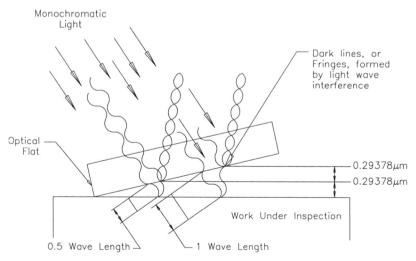

Fig. 7-6. Illustration of the operation of the optical flat and interferometer using the interference of lightwaves to create light and dark bands, or "fringes."

of an inch long, each half a wavelength is roughly 11.6 millionths (0.0000116″ or 0.293 μm). If we refer back to our previous illustration, it can be seen that each dark band will represent an elevation difference in progressions of 0.293 μm.

The optical flat is a quick and simple means of making highly accurate measurements. By resting the instrument directly on the work surface, it is used to check for runout. Placing the instrument half on a workpiece and half on a master gauge, with a short span between them, can be done for checking height. Finally, by placing the workpiece and a master gauge side by side, the optical flat can be used for checking parallelism of the workpiece.

Remember, the optical flat is powdered quartz and, as such, is a fragile instrument much like glass (which it strongly resembles). Always lift and set the instrument very carefully to avoid scratches, chips, or cracks. Keep the surfaces of the instrument clean and free of fingerprints. Due to the very sensitive nature of the measurements being taken, the workpiece should be absolutely clean and free of any sharp burrs. Any material on the workpiece of the contacting surface or the optical flat will cause the instrument to sit unevenly and thus give a false reading.

INTERFEROMETRY

The principles just covered for producing the dark bands with the optical flat are essentially the same principles used in the "interferometer." The interferometer utilizes a stream of a very specific wavelength of light to produce patterns of dark bands. Each dark band is called a "fringe." These patterns of fringes can be much more complex than those typically viewed with an optical flat, and consequently tell a great deal more about the contour of the workpiece surface. Some interferometers have the capability of both viewing the pattern of bands in "real time" on a video screen and photographing the pattern to produce a permanent "interferogram."

Interferometry is another of those highly complicated fields beyond the scope of this book; but an extensive list of further readings on the subject is listed in Appendix B. However, we will cover very briefly some of the principles of interferometry and the contours considered in interferometric patterns. Fortunately,

Fig. 7-7. The interferometer utilizes the same principles as the optical flat, but is capable of much greater detail about the topography of the surface under consideration. (Photo courtesy of Zygo Corp.)

many interferometers also make use of computer hardware and software which allow the instrument to display a three-dimensional image of the contour under inspection. The instrument in Fig. 7-7 illustrates a typical interferometer, while Fig. 7-8 illustrates a noncontact surface profile system using interferometry in conjunction with computer interpretation and display.

During the discussion of the optical flat, the dark bands formed were considered to be straight lines, and the spacing between the lines of no real consequence. With the interferometer, the distance between the lines *is* of great importance, and the lines are only occasionally straight. For example, if the lines are straight (and there are in fact lines appearing), then the surface under view is flat, but tilted in relation to the direction of viewing (surface runout). If, however, the lines appear as concentric circles, then the surface may be straight to the direction of view, but either concave or convex. This is where the distance between lines will tell the operator something about the contour being viewed. Obviously, directly near the top of the crown, the circles will tend to be further apart as the surface curves more toward

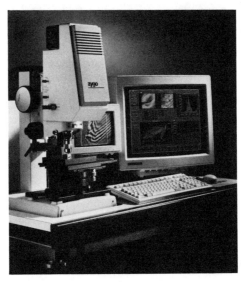

Fig. 7-8. An advancement in the field of interferometers, the three-dimensional surface structure analyzer uses computer assistance to generate three-dimensional images of surfaces on a display screen, relieving the operator of the need to analyze complex interferometric fringe patterns. (Photo courtesy of Zygo Corp.)

perpendicular to the line of viewing. Conversely, the lines will be closer together on the sides of the crowned surface because the tilt away from the direction of viewing will become increasingly greater. Fig. 7-9 illustrates this principle, showing the two-dimensional interferogram and a three-dimensional representation of the surface under consideration.

It is rare, however, that every surface considered will either be only tilted or a nice neat concave or convex. Chances are that a surface under such a high degree of scrutiny will prove to have other flaws similar to those discussed under surface profile indicators mentioned in the last chapter (see Fig. 6-4). These flaws will appear as certain types of aberrations in the line pattern on the display screen of the interferometer. The next easiest type of configuration to visualize is a quasi-torus, where the surface is concave in the center leading up to a convex surface at the outer periphery. In this case, the concentric circles of the interferogram will be tightly grouped near the center, and again further out, with none showing directly at the center or between the two

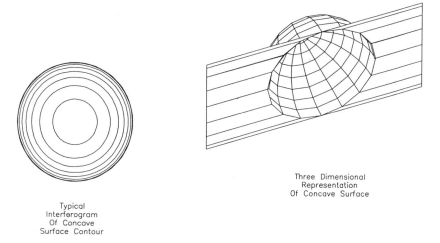

Typical
Interferogram
Of Concave
Surface Contour

Three Dimensional
Representation
Of Concave Surface

Fig. 7-9. Illustration of an interferogram and the three-dimensional surface that is represented by the fringe pattern.

groups. A problem arises, however, if the situation were exactly reversed, i.e., convex in the center and concave further out. This would produce a concentric circle pattern virtually identical to the first case.

The operator must now determine if the pattern represents the former or the latter condition. Although a complete treatment of this procedure is beyond the scope of this book, suffice it to say that the part must be tilted slightly and the direction of the motion of the bands observed. This is largely why the addition of computer imaging equipment makes the use of the interferometer so much easier.

There are many possible patterns of the interferogram. The surface may be a straight cylindrical crown, as though viewing the side of a pipe. This produces a pattern where the fringes are in straight lines but grouped near the outer edges of the interferogram and absent in the middle. Keep in mind, however, that the straightness of these lines is affected by the straightness of the direction of view. A tilted view will cause the lines to curve.

The next surface aberration would be something resembling a potato chip, where the cylindrical curvature happens in two axes. This can produce fringe patterns where the lines appear to flare out and away from each other. There are many other considerations in the use of the interferometer, such as surface formations called a "coma" and "astigmatism," but we will leave these for a more advanced text and to those of you who will be operating these specific instruments.

LASERS

Lasers are a fairly familiar item in modern society, with common applications such as the bar code scanner used at most grocery stores for price registering at checkout. They are also the basis of some of the most sensitive measurement equipment available. The word "laser" is actually an acronym for (L)ight (A)mplification by (S)timulation of (E)mitted (R)adiation. In simplest terms, this means that a light source is amplified into a very intense light "beam," by introducing white light (a combination of all light colors) into a nearly closed chamber containing a pure "lazing" material. This lazing material may be a crystalline solid like ruby or sapphire, a liquid, or even a pure gas such as carbon dioxide. The choice of lazing material depends on the desired laser light to be produced.

The advantage of laser light in measuration is not in the force but in the ability of the beam to be controlled. Ambient light tends to reflect, refract, diffuse, and generally jump all over the place. Laser light, on the other hand, will more or less travel in a constant beam, at a constant width, and at a constant intensity. Intensity is important because we want to measure a part while not damaging it. Many measurement instruments rely on laser light with a very limited range of intensities, and therefore there are measurement instruments which are designed just for measuring the intensity of light emitted from a particular laser. Lasers will also operate on different colors of light, i.e., different wavelengths of light, depending on the application.

Laser light is fairly easy to understand once you have an understanding of light in general. Light has several properties: it travels in waves, the "wavelength" determines the color of the light, the "amplitude" determines the brightness of the light. In laser light the wavelength is determined by the lazing material used in the chamber of the laser generator. When white light is introduced to the laser generator chamber, it is absorbed by the atoms of the lazing material, which leaves them in an elevated state of energy. When the introduced white light is removed, or the atoms have absorbed all of the energy they are able to, the electrons of the lazing material will fall back toward the nucleus of the atom and thus release the absorbed energy. This released energy is in the form of one very specific wavelength of light.

"Frequency" is the number of waves that pass a point in a given period. Generally, this time unit is one second, and the resulting frequency rate is given in cycles (waves) per second, or "hertz." Since frequency is a rate of speed, and the speed of light is a constant, frequency becomes a function of the wavelength. The mathematical function would be $c = \lambda \times f$, where c = the speed of light, λ = the wavelength of a light, and f = the frequency of the light. Obviously, since c is a constant, when λ fluctuates f will fluctuate in the opposite direction.

Lightwaves from ambient light are of various wavelengths, amplitudes, and frequencies. When two lightwaves cross each other, they cancel each other out. The cancellation effect of lightwaves as they cross each other is very valuable, even with lasers, for interferometry. This combination of multiple lightwaves and cancellation makes ambient light of little value in the measuration process.

Laser light, on the other hand, is created so that only one specific wavelength, amplitude, and direction is used. Additionally, all of the waves of this uniform light are synchronous, or what is called "coherent" light.

The laser generator is a nearly sealed chamber, except for one small opening for the laser light to escape from. Typically, the inside of the entire chamber, and always the end opposite the opening, is mirrored to reflect the laser light back into the chamber. In this way, nearly all of the released light energy is eventually channeled out through the opening. This structure is also what is responsible for the coherent nature of laser light.

Because the laser light is of one specific wavelength which is very accurate and consistent, and traveling in virtually a straight line, it will stay in a straight path at a constant speed. These characteristics of laser light are what make it valuable as a measurement tool. Fig. 7-10 illustrates the effects of lightwave cancellation and coherent light.

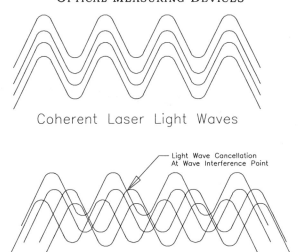

Coherent Laser Light Waves

Incoherent Ambient Light Waves

Fig. 7-10.   Illustration of the cancellation effect of ambient light and the coherent nature of laser light.

Laser Micrometers

The laser micrometer operates similarly to the Vernier micrometer, at least in the respect that it is a linear measuring device. Beyond that similarity, the two instruments start to differ radically. The laser micrometer is a highly accurate noncontact linear measuring device. Because this instrument operates on laser light, rather than physical contact, it is typically accurate to 0.000002 inch. Fig. 7-11 illustrates the typical laser micrometer.

Technically, the laser micrometer doesn't actually measure a part, but rather measures the shadow of the part passed through a laser beam. The laser micrometer operates by interrupting a low-intensity constant laser beam with the part being measured, and registering the gap left in the beam by the part's "shadow."

Obviously, the part being measured would have to be held very square to the instrument given the degree of accuracy it is capable of. For this reason, the laser micrometer lends itself best to measuring diametrical parts. Since it is also a noncontact measuring instrument, it will neither mar a surface by contact nor require the part to leave processing for measuring. Perhaps the best example of this is the application of the laser micrometer to the production of drawn wire, both coated and noncoated.

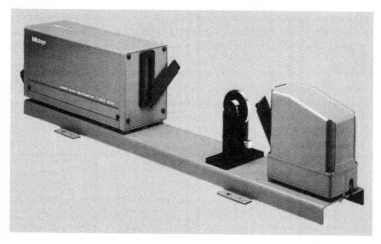

Fig. 7-11. The laser micrometer is a noncontact measuring instrument capable of measuring work in progress without removing it from the production process. (Photo courtesy of Mitutoyo/MTI Corp.)

As the wire is drawn to size it to the proper diameter, it is passed through the measuring field of a laser micrometer to continually check the wire diameter. Again, after coating, the wire is passed through the measuring field of the laser micrometer to check consistency to the size of the coated wire and therefore the proper thickness of the coating.

The real advantage to this arrangement is not so much in knowing if the wire or coating are the correct diameter, but in being able to make instantaneous and constant adjustments to the machinery forming the wire or coating it. Let's just consider the coating process for a moment. As the wire leaves the drawing process, it is passed through a sleeve where molten plastic is injected around it as a coating. The linear travel of the wire is very fast, with hundreds of feet of wire coming out of the draw dies every minute. The coating must be consistent at the same pace.

As the wire leaves the coating extruder, it passes through the measurement field of the laser micrometer, which takes the measurement and relays it to a computer terminal. The computer terminal is programmed with the predetermined diameter of the coated wire. If the value being constantly read by the laser micrometer varies outside of the predetermined parameters, the

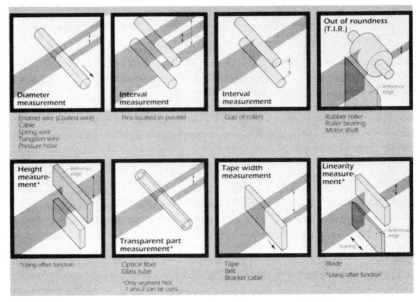

Fig. 7-12. Some typical setups and applications of the laser micrometer. (Illustration courtesy of Mitutoyo/MTI Corp.)

computer will activate another relay which controls the plastic coating flow to increase or decrease its flow. This continual adjustment allows the manufacturer to produce consistently accurately coated wire without the need to stop the line and take periodic readings of the wire diameter, and adjust the coating flow rate. Fig. 7-12 shows other typical setups and applications of the laser micrometer.

Laser Range Finder

We mentioned earlier that many newer transits now use a laser range finder to measure the distance from the transit to the target. This was perhaps one of the first applications of lasers, and certainly one of the first picked up as an application by the military. Using a laser beam to determine distance works on a fairly simple principle, much like that of sonar.

The laser range finder employs a laser generator, a target, an extremely accurate timing mechanism, a computer for converting time information into a calculated distance, and a readout. The range finder fires a laser beam at the target, which in turn simultaneously starts the timing mechanism. The timing mechanism is typically graduated in microseconds (millionths of a second), and some current timing mechanisms are registering time in femtoseconds (quadrillionths of a second)!

Due to the coherent nature of a laser light beam, it remains in a straight beam as it travels to the target, and as it is reflected back. When the beam is reflected back to the initial point of origin, it strikes a sensor on the laser generator and stops the timing mechanism. Since the speed of light, although very fast, is a constant speed, the computer can easily calculate the elapsed time from the moment the laser left until the time it returned, and convert that into a distance using the speed of light in feet per second. The result is displayed on the readout as a distance in feet or meters.

In industrial applications, the target might be a physical object, but it is frequently a "retroreflector mirror," "collimator," or "auto-collimator." Each of these devices works a little differently, but the object is to get the laser beam to reflect back in the same direction that it emanated from, which is less likely with a random workpiece. However, it should be noted that reflection from a workpiece is also a common practice, and very necessary in certain measurement applications such as "Theodolite measuring systems."

Laser Alignment Devices

Laser alignment devices work similarly to the transit, except that the line of sight is replaced with a laser beam. Again, like the line of sight, the laser beam is not subject to sag, motion, or any other interference—it is a perfectly straight line. The transit uses the reflection of ambient light from the target to reach the operator's eye, and the laser uses the reflection of the laser beam back to the point of origin.

Obviously, the laser alignment device uses the same sort of retroreflectors, collimators, sensors, and computer equipment that the laser range finder uses—the difference is how the data from the position of the laser beam are interpreted by the computer software. Once the position of the laser origin is established, it is a matter of shooting it against different targets and collecting the data. One notable application of this device is for digging tunnels. The laser is projected in front of the large tunnel digging machine and the location of the beam target recorded. The machine is then moved forward and any deviation in the position of the digging head is corrected during movement. This technique has proven to keep tunnels straight to an accuracy of a fraction of an inch over several miles of excavation.

Theodolite Measurement Systems

A theodolite is, by definition, a transit. However, in application, the word "theodolite" is associated with a system of three-dimensional measurement similar to the CMM, but using optical instrumentation. The theodolite employs a laser range finder, a laser alignment system similar to a transit, and a sophisticated computer interpretation system to carry out a form of measurement called "triangulation."

To accomplish this triangulation, the theodolite measurement system uses two or more theodolites, each of which locates the same point in space. Typically, that point in space is a point or feature on a physical workpiece. For example, this technique is very popular in the aircraft industry where measuring points on a partially assembled airplane would be virtually impossible without it.

Triangulation is a reasonably simple application of trigonometry. Fig. 7-13 shows the elements of triangulation. Two points, representing the locations of two theodolites, are shown, and the distance between them can be easily found using a range finder. We now have a "baseline" of a known distance. From the line of the baseline, each theodolite is aimed at the same target. Because the theodolite uses two protractors—one for the horizontal (azimuth) and one for the vertical (zenith)—the relative position of the targeted point from the theodolite can be determined.

At this juncture, it is not really necessary to use the range finder to determine the distance from the theodolite to the target. If the target sight point is identical in both theodolites, the lines of sight intersect exactly at the target point, thus forming the third point of a triangle. Each theodolite is equipped with a set of encoders which register the position of the instrument after setting at the baseline. The location of the point where the lines of sight intersect is calculated by the computer and displayed on the computer's screen. Typically, theodolite measurement systems

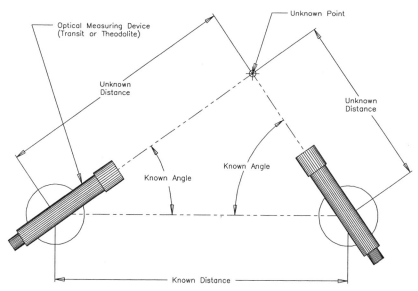

Fig. 7-13. Many optical measuring instruments such as the Theodolite measuring system, use the principle of triangulation for determining measurements. This process replaces linear measurement with angular measurement and mathematics.

will display the location to four places past the decimal point, i.e., to one ten-thousandth of an inch.

Because the angle of the theodolite to the baseline is registered, the computer can use standard trigonometric formulas to determine the length of the remaining sides of the triangle formed by the lines of sight. In this way, linear measurement is being replaced with angular measurement and converted. The protractor head of the theodolite is accurate to one second of arc, so the resulting measurement is very accurate.

Near-Field Laser Microscopy

Near-field laser microscopy is not so much a measurement process as it is an inspection process. It is a relatively new field, specifically when it comes to "scanning near-field optical laser microscopy." The reason I say that this is an inspection process more than a measurement process is primarily due to the resolutions involved. With the toolmaker's microscope, ambient light is used to enlarge an image, e.g., 2 to 200 times. Laser optical microscopy might be used to enlarge an image several thousand times. One recent article published on the subject states, "In this manner, an image is created pixel by pixel with resolution, in some cases, on the atomic level."

We can gather this means that resolutions, and therefore measurements, are in units called "angstroms." The angstrom (symbol Å) is, by definition, equal to one ten-millionth of a millimeter. A look at Appendix A will show you that this means the angstrom is equal to one-tenth of a nanometer; this is substantially smaller than the micron (μ), which is normally considered the shortest unit used in industrial work. However, in all fairness, industry is quite a broad term, and it is responsible for some very sophisticated products like artificial hearts, computer microchips, and concave reflector telescope mirrors measuring 8.5 m (27.89 ft!) in diameter. For these reasons, near-field scanning laser optical microscopy deserves at least a cursory examination in this text.

Optical microscopy is exactly the process discussed earlier with the toolmaker's microscope. It involves simply reflecting a light off an object and channeling that light through a series of lenses to enlarge the image of the object. Remember that this process differs slightly from the normal laboratory microscope in that the light comes from above and is reflected back to the lens rather than passing through a specimen to the lens. The problem with this is that there is a point where the object cannot be enlarged any more. This is called the Abbé barrier, or diffraction limit. In simplest terms, this means that the resolution limit of the optical microscope (in contrast to the electron or ion microscope) can be no greater than one wavelength of light, or about 500 μm, in general practice.

To overcome this barrier, the light source is supplied very near to the object being observed (whence the name "near field") via an opaque metallic covered fiber optic filament. The fiber optic filament must be maintained at a distance of 5–10 nm from the surface of the observed object to maintain proper resolution. This means that the fiber optic is literally following the topography of the surface, which in turn provides the operator with a topographical map of the surface simultaneously with a visual image. This distance is maintained by a sophisticated means of force detection to electronically adjust the position of the fiber optic tip.

Scanning near-field optical laser microscopy is finding application in biomedical research, and in the field of microelectronics and semiconductors. It is likely that from a purely industrial point of view, this instrument will find increasing use in the field of semiconductors, where measurements are being used which are now in tenths of microns.

While the advent of superresolution electron and ion microscopes has changed the way scientists and engineers view many things, there are some obvious advantages to optical microscopy, and especially scanning near-field optical laser microscopy. Optical microscopy provides an image that is much more "normal" in the sense of being the sort of image the human eye is used to seeing. A close analogy might be that color infrared photography has some invaluable applications and can reveal things standard photography cannot; but interpreting those images requires a new way of looking at them and a lot of training. Another advantage, and perhaps even more important, is that the process necessary for electron microscopy requires the specimen to be prepared by elaborate means. Optical microscopy, on the other hand, can view a specimen with little or no preparation. In essence, this means that optical microscopy is a nondestructive, or nonintrusive, means of viewing an object, whereas electron microscopy typically places the object in a foreign and generally destructive environment.

Machine Tool Analyzer

Many of the principles covered so far are frequently combined into specialty measuring machines. A good example of this practice is the "machine tool analyzer" pictured in Fig. 7-14. This instrument is used for measuring all of the features of machine cutting tools, including: drills, reamers, milling cutters, ball-end cutters, taps, and the like. The tool bit is placed in a chuck or collet, positioned using linear and angular Vernier micrometers, and sighted with a microscope. The microscope is mounted similarly to a transit, and the objective lens employs a reticle pattern similar to both for measurements.

In Fig. 7-15 we see a closeup of the tool being sighted with the tool analyzer, and the reticle pattern superimposed in the background. The collet is mounted in an angular Vernier scale for

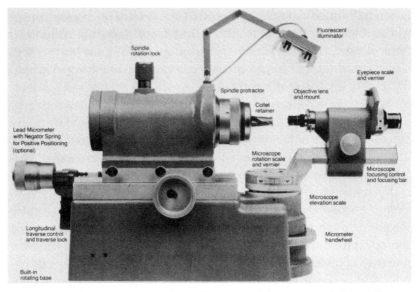

Fig. 7-14. The machine tool analyzer employs many of the different principles of measuration in one integrated, specialized measuring instrument. (Photo courtesy of Stocker & Yale, Inc.)

Fig. 7-15. The machine tool analyzer utilizes a reticle pattern similar to those found in transits, optical comparators, and measuring microscopes for measuring virtually every feature of machine tool cutting bits. (Photo courtesy of Stocker & Yale, Inc.)

positioning the tool bit, while the entire collet is positioned using a linear micrometer attached to a slide table. The sighting micro-scope is also adjustable in all axes, with a rotational Vernier around the axis of the scope, another angular Vernier at 90° to that axis shown in the bottom center of the picture. The scope is also adjustable in/out from the tool bit.

The machine tool analyzer will employ this combination of mechanical and optical measuring devices to measure distances and angles of every type on the tool bit. The angle of a drill point or the draft angle of the flutes of a mill cutter, the length of the threads on a tap or the taper lead at the end of the tap are all measurements that can be taken with the machine tool analyzer.

Chapter 8

Miscellaneous Measuring Devices

UP TO THIS POINT, we have described and discussed many of the measuration and calibration instruments used in industry. There are, however, a number of ancillary devices which are either measuration instruments, or are directly related to measuring practices. The latter of these devices can be thought of as devices which either facilitate the use of other measuration devices or act as a reference tool or benchmark. This chapter will cover many of these ancillary devices and their applications to the measuration process.

Keep in mind that we have dealt primarily with the types of measuration instruments common to the manufacturing of solid workpieces, as opposed to the manufacturing of fluid compounds. A complete treatment of measuring would cover an entire wall in most major libraries. In addition to linear measurements, it would include volume, density, viscosity, temperature, elevation, speed, voltage, amperage, wattage, power, pressure, time—the list is nearly endless. The devices covered in this chapter will deal primarily with linear dimensions or the function of industrial machinery.

Ultrasonic Measuring Devices

Ultrasonic measuring is extremely advantageous because it is a completely nondestructive and nonmarring test. It also serves as both a measurement technique and an inspection technique. It works on the same principle as sonar, and in that respect is similar to the laser range finder. In simplest terms, it is like thumping a melon at the market to make a determination about the contents of the inside without opening the melon. For years prior to the advent of sophisticated ultrasonic measurement devices, machinists would strike a workpiece or weldment with a hammer and listen for the clarity of the tone to make a judgement about the quality of a weld. Now with the aid of electronics, the sound waves can make precise readings of part thicknesses and composition.

The theory of sonic, or ultrasonic, inspection is based on the fact that sound travels at a constant speed through any given material. The speed will also vary as it travels through different materials, or similar materials of different composition. For example, the speed of sound in air is 1087.1 feet per second. However, that speed is only correct in dry air at 0°C, at sea level. If any of these criteria change, the speed factor will change. Therefore, sonic measurement and inspection is dependent on knowing the exact composition and condition of the material being inspected.

The ultrasonic measurement device uses a transmitter/receiver probe placed against the workpiece, which emits an ultrasonic (above the range of human hearing, i.e., beyond 20,000 Hz) sound or "ping." The sound wave travels through the material of the workpiece, and when it reaches the far side of the workpiece, part of the sound is reflected back to the receiver of the probe. Various instruments will then make different use of the information received. Some use a monitor screen that is a type of "oscilloscope," while others will internally calculate the elapsed time to make a measurement determination.

Fishermen use a similar apparatus to make determinations about the water under them. This device will generate a picture of the contour of the bottom of the body of water, and simultaneously indicate obstructions, which at least ostensively are fish. The industrial ultrasonic measurement device does more or less the same thing, but in a sort of reverse sense. As the sound wave passes through the workpiece, it will travel faster through trapped pockets of air in welds. These pockets represent flaws in the weld and can be displayed quite accurately on a screen without the need to cut the weld in half to view the flaw.

Another advantage of ultrasonic measurement is in those circumstances where the far side of the measurement to be made is not accessible. This might be any sort of enclosure, such as hollow casting or injection molded plastics, or drawn or extruded shapes of such length that accessing the inside of the part is impossible.

X-Ray Measuring Devices

Over the course of the last several years, I have had the occasion to have all or part of my head examined with X-rays (no joke), which gave me a special insight into the use of this technique for industrial inspection. Any process used with sufficient delicacy to be functional on the human head is certainly a good candidate for nondestructive testing on industrial workpieces.

There are two types of X-ray inspection devices used in industry. The standard X-ray works just like the medical X-ray, whereby X-rays are passed through the part being inspected and onto a special film. The resultant X-ray photograph can then be inspected to make a determination about the internal structure of the workpiece. The other type of X-ray inspection device employs a specially coated fluorescent display screen for displaying the image in real time. This latter devise is aptly called a "fluoroscope."

In either case, the image is produced as dark areas on a light background. The X-ray measurement utilizes a part of the electromagnetic spectrum with very short wavelengths. In this respect, it qualifies as an optical measuring device; but due to the special, and to a certain degree inherently dangerous, nature of X-rays, they are categorized by themselves. The images of X-ray inspection are essentially a black and white image caused by greater or lesser amounts of absorption of the X-rays as they pass through or are reflected back from the workpiece.

Some of the typical applications of X-ray inspection are for checking the integrity of pressure vessels, and for several railroad applications. For example, the wheels of railroad locomotives and cars are cast in a spin-cast process. This causes the impurities in the metal to flow to the outer face of the wheel. Once the wheel is placed in operation, these impurities work-harden, thus increasing the durability of the wheel. However, X-ray inspection is a good means of determining if flaws or blow-holes have formed in the wheel during manufacture. Another of the applications of X-ray inspection to railroad operations is the use of periodic fluoroscopic inspection of rails in place to detect flaws that might develop during use. Obviously this is a situation where inspecting the workpiece out of position would be, to say the least, inconvenient. The flaws can be detected and a determination can be made whether they are large enough to merit replacement.

Rotational Speed Indicator

Although not a linear measurement device, the rotational speed indicator is an important measurement tool in industry. The instrument vaguely resembles a manual speed drill, but without the crank handle. In place of where the handle would be on a speed drill is a dial with graduations. The spindle which extends from one end rotates freely in a sealed, lubricated bearing. This spindle is placed in or against a rotating spindle, and the internal gear mechanism registers the revolutions of the spindle on the side dial.

Since the rotational speed indicator only registers the number of revolutions while being turned by the machine spindle, it is necessary to time the engagement time of this instrument. For example, if the indicator is engaged to a machine spindle for one minute, the resulting number of revolutions registered on the dial will automatically be the RPM's (revolutions per minute) of the spindle.

If the motion in question does not provide an easy means of engaging the spindle of the instrument, a rubber-coated wheel attachment is available which fits on the spindle of the instrument. However, to compensate for the larger diameter of the wheel, the number of revolutions registered must be divided by 2 before RPM's are calculated.

The correct rotational speed of spindles, wheels, and shafts is frequently an important consideration in machine operation. Frequently, spindle speed is the cutting speed of metal-removal machinery, or shaft speed must be maintained on many special machines because operations are "timed" to happen in a very precise sequence. The rotational speed indicator allows the operator or machine installer to regulate and adjust machine operations for proper functioning.

Material Hardness Testers

While all sorts of tests are constantly needed on the various strengths of materials, hardness testing is perhaps the most commonly performed and most necessary. Tensile strength, flexural modulus, shear load, and torsional strength are all common and necessary tests, but hardness is more so because the wear resistance and performance of materials is typically of the most concern. Material hardness testers determine the hardness of a material sample as a function of the amount a probe will penetrate the material at a given force load.

The hardness of material is expressed in terms of a particular value on a scale that reflects the penetration of the type of probe for the given force. The most common of the scales are the "Rockwell Hardness Scale" and the "Durometer Scale." Generally, the first scale is used for harder materials like metals, and the latter scale is used for softer materials like plastics and rubbers. There are a number of other hardness scales such as "Brinell," "Vickers," and "Scleroscope Hardness Test," which will appear in engineering drawing standards from time to time. Each has a slightly different way of determining the hardness of materials; but most do so in a similar manner, i.e., by impacting the specimen with an object and noting the reaction.

Fig. 8-1 is a typical Rockwell hardness tester. The Rockwell hardness test can be subdivided into four other scales, each designated by a letter: A, B, C, and D. The A and D scales are the least commonly used, as they are specialty scales for thin specimens and case-hardened materials. Note also that, depending on the age and manufacturer of the instrument, these scales may be designated N and T. The C scale is the most common, and is used

Fig. 8-1. The Rockwell hardness tester measures the hardness of materials. These values will tell the operator about other characteristics of the material as well. (Photo courtesy of Mitutoyo/MTI Corp.)

for most hard materials such as steels. It uses a 120° inclusive angle pyramidal probe point. Finally, the B scale is used for softer metals such as cast iron and nonferrous metals, and uses a 1/16 diameter ball probe. Additionally, each scale uses a different set of forces for determining hardness. Each scale is represented separately on the dial of the Rockwell hardness tester.

The Rockwell C scale test is performed by placing the specimen on the anvil and applying a 22 lb (10 kg) load against the probe. The dial on the instrument is then zeroed, and then the load is increased to 330 lb (150 kg). The load is now relieved and the hardness read directly from the C scale on the instrument. This value is an interpretation of the amount of penetration of the probe into the material.

Brinell hardness testing is similar in the respect that a probe is forced into the surface of the specimen. A 10-mm-diameter ball probe is forced into the specimen under a 3000 kg load, which leaves a concave impression. To obtain the hardness, the diameter of the impression is measured with a microscope. Obviously the process is somewhat more laborious than the Rockwell test.

The Vickers hardness test is similar to the Brinell test, except that it uses a pyramidal probe to produce a diamond impression in the specimen. The diamond impression is then measured across the points with a microscope and the distance compared to a chart to obtain the hardness rating.

The scleroscope hardness test operates just a little differently. A probe of a certain weight is actually dropped on the specimen, and the amount it bounces is recorded by the instrument. There are some advantages to this particular test: very thin parts can be tested which would be completely penetrated by a test such as the Rockwell test; the test can be performed comparatively quickly, allowing for a greater number of samplings to be done in the same amount of time; and very hard parts which would not test well with other methods can be tested with ease.

Level

There are essentially four geometric conditions of concern to builders and machinists: flatness, squareness, levelness, and plumbness. We have covered a number of the ways levelness is determined, both electronically and optically, but the simplest and most commonly used is a manual level. The level is essentially the spirit level we discussed in Chapter 7 under "Transits," but now we are referring to a level mounted in a long frame with precisely parallel sides.

The spirit level is a sealed glass tube, curved in the plane of the parallel side of the level, and filled with sufficient fluid to leave a bubble. The bubble, you will recall, floats up along the curvature of the tube away from gravity. This curved vial is fixed in the frame of the level so that the ends of the vial are in a line parallel to the surface of the level frame. In this way, when the frame is set on a surface, the bubble of the spirit level will run true to the surface the level in sitting on.

Levels are available in a variety of lengths, from a few inches long to six or eight feet long. There are some considerations concerning the length of a particular level. The level will only sit on the two highest points of the line it is set on. Therefore, the levelness determined by this instrument is determined by those two points. Usually, the surface being checked is straight, but this is not always the case, and a longer level will make these cases more difficult. For example, if the surface being checked is a structural wood 2 × 4 that has some warpage, determining the correct position of the level along the crown of the warp is more difficult as the level becomes longer.

Conversely, since the level does sit between only two points, the longer level will give a more accurate reading if the surface is reasonably straight. Levelness is determined in terms of its counterpart, "runout." Runout is expressed in terms of distance of deviation from levelness per distance of linear travel. This might be something like 1/16 inch per foot, which means the surface moves out of perfect level 1/16 inch for every 1 foot of horizontal measurement. Making this sort of determination is easier to do with a somewhat longer level. The reason should be fairly obvious. If the amount of runout expressed in our last example was measured with a 6-inch-long level, the runout over the course of the entire instrument would only be 1/32 inch, and could be conceivably missed entirely. If, however, the level used were 8 feet long, the total runout over the entire length of the instrument would be 1/2 inch, and therefore a very visibly obvious error.

Plumb Bob

The "plumb bob" is one of the simpler measurement instruments. It consists of a short hardened steel cylinder, about 3–4 inches long and about 1 inch to 1-1/2 inches in diameter, with one pointed end and a place to attach a string at the other end. The plumb bob is so named because it is used to ascertain "plumbness," which is a line perpendicular to the horizon. It is simply a weight of exact balance around a central axis used to hold a string taut in the direction of gravity.

The plumb bob is a rather crude instrument in relation to the high-technology devices we covered in the last chapter, but it is fairly accurate and has several advantages. It is very cheap, very fast, very durable, and the operator can carry it virtually anywhere in his pocket. We also mentioned that the plumb bob is in fact used to establish the center location of the transit stand for transit operations. For operations such as residential construction, the plumb bob is used to make sure the framing of walls is straight up and down. A transit would do this, but it is unnecessarily accurate, and much slower to set up. In this case, the more accurate instrument is prohibitively expensive just in terms of the man hours to operate it.

Ancillary Tools

The remaining few sections deal with instruments which don't actually make measurements, but are necessary in many of the "layout" operations where measurements are made, or instruments used to locate those measurements once they are made for the machining process. Layout is a process where measurements taken from engineering drawings are transferred to pieces of metal or other materials for final machining. This commonly involves scribing lines on the metal to follow with a machine cutter. This is the first group of ancillary tools.

The second group of ancillary tools comprises instruments used in conjunction with machines for locating those scribed lines before machining begins. In this respect, it is obvious that these tools are never used by themselves, they are in fact an aid to the machine doing the processing. These instruments act as an interface between the measurements being made and the process they are made for to bring the two parts of the operation into synchrony.

SCRIBERS, CENTER PUNCHES, AND BLUING

When a machinist is presented with an engineering drawing of a part to be made, he is faced with the task of transferring that information onto a piece of material. Typically that material is metal, and most frequently steel. We will proceed on the assumption that steel is our material of choice in the remaining example, because the operations will work the same regardless of material.

Transferring the lines of the engineering drawing onto steel requires the machinist to draw or otherwise scribe those lines onto the metal. This must also be done in such a fashion so that the lines are easily visible and comparatively stable, i.e., they cannot be easily rubbed off. Since drawing on metal with pen or pencil doesn't work very well, and soapstone is large and therefore inaccurate, the machinist will typically scratch, or "scribe," the information necessary onto the metal with a special hardened "scriber."

The scriber is a pencil-sized piece of alloy steel specially hardened to scratch a line in steel without suffering any excessive wear itself. The end of the scriber is ground to a fine point and produces a very fine, thin line which rivals or exceeds any produced by a writing instrument. Because of both the sharpness and the hardness of this instrument, the operator need not apply any undue force to produce an acceptable line.

However, even though a line scratched into the metal is visible and will not rub off, it still can be fairly difficult to see if it is exactly the same color as the background material. To alleviate this problem, machinists will coat the metal with a dark blue dye simply called "bluing." Bluing is a dye dissolved in a ketone solvent which dries very rapidly. Usually drying is complete in less than one minute. Once the bluing is applied to the surface of the metal to be laid out, the machinist can scribe the necessary lines through the bluing, which leaves bright silver lines on a blue background. This makes it very easy to read during the machining process even when the workpiece is cluttered with metal chips. Once machining is complete, the remaining bluing is removed with acetone or other special solvent. Fig. 8-2 illustrates a machinist scribing lines for machining on a sheet of blued metal.

Some machining operations require an additional operation before machining can begin. For example, drilling into a flat piece of metal is not as easy as it might sound, especially if the drill bit is not absolutely sharp. The drill will tend to wander away from the point it touches until it begins to cut into the metal. This condition worsens as bit length increases, but even a small amount of wander can cause the hole location to be out of position beyond acceptable tolerances. While it is easy enough to bring the drill bit down onto a set of scribed crosshairs, it is not always possible to tell if the drill wandered away from the intersection of the lines before cutting them out.

To solve this problem, a small pointed "center punch" is placed at the intersection of the appropriate scribed lines and driven slightly into the metal with a hammer. Crude, but effective, this leaves a small pocket exactly at the intersection of the cross-

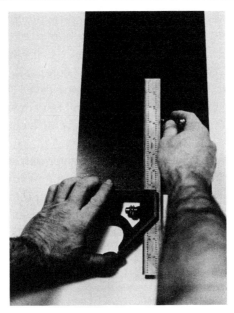

Fig. 8-2. Machinists commonly use bluing to enhance the visibility of lines scribed at specific measurements. This is the most frequent method of transferring measurements to the workpiece. (Photo courtesy of The L.S. Starrett Co.)

hairs that the drill bit point will now seat in and not be allowed to wander.

One more word is in order about scribing straight lines on metal. Frequently, a special scriber is attached to another instrument, as is common with the use of a height gauge. The blued part is set on a surface plate, and the height gauge is set to an exact height and the scriber dragged along the metal. Scribing layout operations are frequently carried out on a surface plate to assure accuracy. Neither the bluing nor the acetone solvent will damage the surface plate, but the plate itself should always be kept free of any spilled bluing.

TRAMMELS

Trammels are another of the layout instruments used in conjunction with bluing. They consist of a set of two scribers, each set into a frame that is fitted onto a long bar. This instrument is, in essence, a machinist's compass, or, more properly, a beam compass, because it is suspended by a beam much like a draftsman's beam compass.

Again, the metal is blued, and the necessary lines are scribed into the bluing. Since this is a type of compass, the center, or vertex, of the arc to be scribed is laid out first. Then one of the trammel points is placed at the vertex and held down firmly, while the remaining point is dragged along to create a scribed arc in the bluing.

Trammels have certain advantages over using a radius template to scribe around. First, when scribing around a template, the thickness of the scribe point must be taken into consideration to obtain an accurate arc. With trammels, the scribe radius can be scribed on the metal as a small tick and then extended with the trammels, or the trammels can be set to the appropriate radius directly on a scale.

CENTER FINDER OR "WIGGLER"

Let's assume that the machinist is going to drill a large hole in a steel plate. Unfortunately, the diameter of the drill makes visually aligning the point of the drill bit with the scribed center point of the hole location very difficult or impossible. In this situation, the machinist will probably resort to a two-step process, using a center finder or "wiggler."

The center finder is a needle-like probe mounted on a ball socket in a spindle. The spindle mounts in the chuck jaws of a machine like a drill press or milling machine. When the spindle of the machine is engaged, the center finder's probe will revolve around its center of gravity. Because of the nature of the motion, if a blade (e.g., that of a T-square) is brought up against the needle probe, the needle will come into line with the axis of the machine's motion.

The machinist now has a needle probe point pointing directly in line with the axis of the spindle. The table of the machine is then traversed so the probe point of the center finder is directly in line with the intersection of crosshairs of the hole location. Once in place, the table is locked in position for both horizontal axes and the machine is turned off. The table is dropped in the vertical axis, and the center finder removed and replaced with the appropriate drill bit. The machine can now be turned back on and the hole drilled exactly in position. The center finder is one of the instruments we mentioned earlier in this chapter as serving as the interface between making a measurement and using the measurement.

EDGE FINDER

One of the typical applications of the dial indicator is to set the sides of a workpiece parallel to the direction of travel of a machine table. In this way, features can be machined into the workpiece using straight line Cartesian coordinates with the straight line motion of the machine table. However, once the workpiece is set parallel to the motion of the table, the location of the machine spindle to the workpiece must be determined. Frequently, dimensions are taken from a machined edge rather than from a machined hole, so that the spindle must be set relative to the machined edge of the workpiece. This is accomplished with an "edge finder."

The edge finder is an instrument similar in operation to the center finder described in the last section. It consists of a shank with a floating end piece which rotates exactly around the center of the axis of the shank. The end piece can float off center approximately 1/32 inch. The contact shaft area is typically 0.200 inch in diameter. Fig. 8-3 is a typical edge finder.

The instrument is placed in the chuck or collet of the machine, and rotated, as the machine table is traversed until the edge finder contacts the side of the workpiece. When the floating spindle rotates concentric with the axis of the machine spindle, the axis of the machine spindle is now one-half of the diameter of

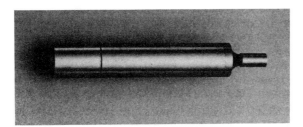

Fig. 8-3.   The edge finder is used to zero a machine table and workpiece in relation to the spindle of the machine. (Photo courtesy of The DoALL Co., Des Plaines, IL.)

the edge finder contact surface from the part edge. This means that the center line of the machine spindle axis is exactly 0.100 inch from the edge of the part.

The dials of the machine table can now be zeroed, and the 0.100 inch subtracted from the reading to set the position of the spindle relative to the part—it is subtracted rather than added because it is assumed that zero is at the part edge and will progress away from that point. Naturally, different situations will dictate whether the 0.100 is added or subtracted. This instrument will normally give a true distance to the edge of the part within 0.0005 inch. Machining can then proceed to each distance, indicated on the engineering drawing, from the machined edge.

Epilogue

This is where the author gets the opportunity to speculate, and to invite the student intending to make a career of the science of measurement to join in that speculation. Some of this may sound like it borders on science fiction, but many scientific developments have arisen from what was once science fiction. Jules Verne's vision of deep sea and space travel was science fiction in his day—now it is reality.

So what does the future hold for advancements in the field of measurement? We have seen that many of the advancements in measurement have tended toward the use of light, in the form of using the lightwave as a standard of measurement, and the use of lasers as measuring tools. Electronic leveling devices and computers have changed the art of measurement in recent years. These devices, no matter how sophisticated they appear now, will one day be as archaic as the cubit (the distance from the tip of a finger to the elbow). Lightwaves will be considered too erratic to use as a standard.

The student of measurement will probably be looking at measurement systems controlled by computer sensing devices based on patterns associated with the "tachyon," "quark," or some antimatter particle such as the "positron." The positron is a particle of equal mass to an electron, but with a positive charge, and thus considered an "antimatter" material. Typically, these currently are only thought to exist briefly, at least in our dimension, at the center of a nuclear reaction. The quark is at least partially theoretical at present. It is considered to be one of the subatomic, or sub-subatomic, particles that comprise things like mesons and photons. The light devices we currently use obviously employ the photon, so using the quark in the future is not that much of a stretch of the imagination.

Using the tachyon as a standard of measurement is still a long way off. This "particle," if it can truly be called that, is a completely theoretical entity which travels faster than the speed of light. Can't be done, you say? Why not? If we accept that matter does not convert to energy until it is accelerated to the "square" of the speed of light—and who are we to argue this point with Einstein—then there must be the possibility of speeds for matter between light speed (186,282 miles per second) and its square (34,700,983,524 miles per second). Light travels in waves theoretically because space travels in waves, according to Minkowsky, but the tachyon may not travel in waves depending on whether it is bound to "real space." If it does travel in real space, and in waves, these waves may become the new standard of measurement. If it doesn't, perhaps there are other means of using its motion or size as a standard of measurement.

English/Metric and Metric/English Conversions

Table of Metric Prefixes

Metric Prefix	Symbol	Multiples and Submultiples
Tera	T	1,000,000,000,000
Giga	G	1,000,000,000
Mega	M	1,000,000
Kilo	k	1,000
Hecto	h	100
Deca	da	10
Deci	d	0.1
Centi	c	0.01
Milli	m	0.001
Micro	μ	0.000001
Nano	n	0.000000001
Pico	p	0.000000000001
Femto	f	0.000000000000001
Atto	a	0.000000000000000001

Other Units and Constants

Angstrom (Å)	= one ten-millionth of a millimeter
Mill	= one one-thousandth of an inch (0.001)

Speed of sound in dry air @ 0°C	= 1087.1 ft/sec
Speed of light in a vacuum	= 186,282 miles/sec
Speed of light squared	= 34,700,983,524 miles/sec

Wavelength of orange-red light of krypton 86 = 6057.802 Å

Inch	= 42,016.807 wavelengths monochromatic light from Krypton-86
Meter	= 1,650,763.73 wavelengths monochromatic light from Krypton-86

International Nautical Mile	= 1.15078 statute miles
International Knot	= 1.1508 statute miles/hr
	= 101.269 ft/min
	= 1.6878 ft/sec

Linear Conversions

ENGLISH TO METRIC CONVERSION

Multiply	By	To Obtain
Feet	0.3048	Meters
Feet	0.0003048	Kilometers
Inches	0.0254	Meters
Inches	25.4	Millimeters
Yards	0.9144	Meters
Miles	1.609344	Kilometers
Mills	25.4	Microns
Square inches	6.4516	Square centimeters
Square feet	0.09290304	Square meters
Square yards	0.83612736	Square yards
Cubic inches	16.387064	Cubic centimeters
Cubic feet	0.02831684659	Cubic meters
Cubic yards	0.764554858	Cubic meters

METRIC TO ENGLISH CONVERSION

Multiply	By	To Obtain
Meters	3.280839895	Feet
Meters	1.093613298	Yards
Millimeters	0.03937007874	Inches
Centimeters	0.3937007874	Inches
Kilometers	0.6213711922	Miles
Square centimeters	0.1550003100	Square inches
Square meters	10.76391042	Square feet
Square meters	1.195990046	Square yards
Cubic millimeters	0.06102374409	Cubic inches
Cubic meters	35.31466672	Cubic feet
Cubic meters	1.307950619	Cubic yards

Angularity

1 degree	= 0.01745 radians
1 degree	= 1.11111111 grads
1 radian	= 57.2958 degrees
1 radian	= 63.66197724 grads
1 grad	= 0.90000000 degrees
1 grad	= 0.01570796 radians

Suggested Further Reading and References

C.W. Kennedy, E.G. Hoffman, and S.D. Bond, *Inspection and Gaging* (Industrial Press, New York, 1987).

F. Wheeler, *The Sizes of Things* (Coward-McCann, New York, 1968).

H.M. Muncheryan, *Principles & Practices of Laser Technology* (TAB Books, Blue Ridge Summit, PA, 1983).

"Tools & Rules," The L.S. Starrett Co., Athol, MA, 1992.

"Optical Alignment Manual," Cubic Precision, Teterboro, NJ, 1986.

"Interferogram Interpretation and Evaluation Handbook," Zygo Corp., Middlefield, CT, 1977.

D. Malacara, *Optical Shop Testing* (John Wiley & Sons, New York, 1978).

P. Moyer and T. Van Slambrouck, "Near-Field Optical Microscopes Break the Diffraction Limit," *Laser Focus World*, October 1993, pp. 105–109.

Gear Tooth Measurement

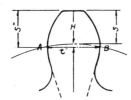

CHORDAL THICKNESS OF GEAR TEETH BASES OF 1 DIAMETRAL PITCH

S = Module or addendum, or distance from top to pitch line of tooth

s'' = Corrected $S = H + S$

t'' = Chordal thickness of tooth

H = Height of arc

When using gear tooth Vernier caliper to measure coarse pitch gear teeth, the chordal thickness t'' must be known, since t'' is less than the regular thickness AB measured on the pitch line. In referring to the table below, note that height of arc H has been added to the addendum S, the corrected figures to use being found in column s''.

For any other pitch, divide figures in table by the required pitch.

No. of Teeth	t″	s″	No. of Teeth	t″	s″	No. of Teeth	t″	s″
6	1.5529	1.1022	51	1.5706	1.0121	96	1.5707	1.0064
7	1.5568	1.0873	52	1.5706	1.0119	97	1.5707	1.0064
8	1.5607	1.0769	53	1.5706	1.0117	98	1.5707	1.0063
9	1.5628	1.0684	54	1.5706	1.0114	99	1.5707	1.0062
10	1.5643	1.0616	55	1.5706	1.0112	100	1.5707	1.0061
11	1.5654	1.0559	56	1.5706	1.0110	101	1.5707	1.0061
12	1.5663	1.0514	57	1.5706	1.0108	102	1.5707	1.0060
13	1.5670	1.0474	58	1.5706	1.0106	103	1.5707	1.0060
14	1.5675	1.0440	59	1.5706	1.0105	104	1.5707	1.0059
15	1.5679	1.0411	60	1.5706	1.0102	105	1.5707	1.0059
16	1.5683	1.0385	61	1.5706	1.0101	106	1.5707	1.0058
17	1.5686	1.0362	62	1.5706	1.0100	107	1.5707	1.0058
18	1.5688	1.0342	63	1.5706	1.0098	108	1.5707	1.0057
19	1.5690	1.0324	64	1.5706	1.0097	109	1.5707	1.0057
20	1.5692	1.0308	65	1.5706	1.0095	110	1.5707	1.0056
21	1.5694	1.0294	66	1.5706	1.0094	111	1.5707	1.0056
22	1.5695	1.0281	67	1.5706	1.0092	112	1.5707	1.0055
23	1.5696	1.0268	68	1.5706	1.0091	113	1.5707	1.0055
24	1.5697	1.0257	69	1.5707	1.0090	114	1.5707	1.0054
25	1.5698	1.0247	70	1.5707	1.0088	115	1.5707	1.0054
26	1.5698	1.0237	71	1.5707	1.0087	116	1.5707	1.0053
27	1.5699	1.0228	72	1.5707	1.0086	117	1.5707	1.0053
28	1.5700	1.0220	73	1.5707	1.0085	118	1.5707	1.0053
29	1.5700	1.0213	74	1.5707	1.0084	119	1.5707	1.0052
30	1.5701	1.0208	75	1.5707	1.0083	120	1.5707	1.0052
31	1.5701	1.0199	76	1.5707	1.0081	121	1.5707	1.0051
32	1.5702	1.0193	77	1.5707	1.0080	122	1.5707	1.0051
33	1.5702	1.0187	78	1.5707	1.0079	123	1.5707	1.0050
34	1.5702	1.0181	79	1.5707	1.0078	124	1.5707	1.0050
35	1.5702	1.0176	80	1.5707	1.0077	125	1.5707	1.0049
36	1.5703	1.0171	81	1.5707	1.0076	126	1.5707	1.0049
37	1.5703	1.0167	82	1.5707	1.0075	127	1.5707	1.0049
38	1.5703	1.0162	83	1.5707	1.0074	128	1.5707	1.0048
39	1.5704	1.0158	84	1.5707	1.0074	129	1.5707	1.0048
40	1.5704	1.0154	85	1.5707	1.0073	130	1.5707	1.0047
41	1.5704	1.0150	86	1.5707	1.0072	131	1.5708	1.0047
42	1.5704	1.0147	87	1.5707	1.0071	132	1.5708	1.0047
43	1.5705	1.0143	88	1.5707	1.0070	133	1.5708	1.0047
44	1.5705	1.0140	89	1.5707	1.0069	134	1.5708	1.0046
45	1.5705	1.0137	90	1.5707	1.0068	135	1.5708	1.0046
46	1.5705	1.0134	91	1.5707	1.0068	150	1.5708	1.0045
47	1.5705	1.0131	92	1.5707	1.0067	250	1.5708	1.0025
48	1.5705	1.0129	93	1.5707	1.0067	Rack	1.5708	1.0000
49	1.5705	1.0126	94	1.5707	1.0066			
50	1.5705	1.0123	95	1.5707	1.0065			

Courtesy of the L.S. Starrett Co.

Index